Advanced Practical Chemistry

Edited by

Alec Thompson MA
Advisory Teacher, ILEA

Lambros Atteshlis BSc
Formerly Advisory Teacher
and member of the Science
Support Team, ILEA

John Murray

First published 1985
by John Murray (Publishers) Ltd
50 Albemarle Street
London W1X 4BD

Reprinted 1989, 1990, 1992, 1994

Printed and bound in Great Britain by
St Edmundsbury Press Ltd, Bury St Edmunds, Suffolk

British Library Cataloguing in Publication Data

Advanced practical chemistry
  1. Chemistry
  I. Thompson, A.   II. Atteshlis, L.
  540   QD33

ISBN 0 7195 4188 3

PREFACE

The experiments in this book are those developed as part of the Independent
Learning Project for Advanced Chemistry (ILPAC) produced by an Inner London
Education Authority team. As with all the ILPAC materials, these were tested
in schools and colleges inside and outside London and cover the requirements
of all the major Examination Boards.

An important feature of the complete ILPAC scheme is that it enables students to
work more effectively on their own and at their own pace. This is reflected in
the way in which the procedures for the experiments are written: the style is
personal and direct, and the instructions are far more detailed and precise than
is usually the case in practical books.

However, the phrase 'independent learning' must not be interpreted too literally.
It is not intended that students should be left unsupervised while they do
practical work: rather, it is envisaged that a single teacher will be better
able to supervise a number of different experiments going on at the same time
without having to spell out detailed instructions to everyone.

An important feature of this book is the provision of specimen results, with
detailed calculations where appropriate, and answers to the 'consolidation'
questions which follow each experiment. These are available for use by teachers
in the Advanced Practical Chemistry Resource Pack.

LONDON 1985

ACKNOWLEDGEMENTS

Our thanks are due to the University of London Entrance and Schools Examination
Council for permission to use experiments which appeared in practical
examinations:

Experiment 38 (1980), Experiment 60 (1979), Experiment 61 (1974),
Experiment 67 (1973 & 1980), Experiment 75 (1980), Experiment 80 (1977),
Experiment 88 (1978 & 1979), Experiment 95 (1978 & 1982)

Experiment 44 is based, with the kind permission of Longman Group Ltd, on an
experiment which appeared in 'An Experimental Introduction to Reaction Kinetics'
by M.A. Atherton & J.K. Lawrence (0582 32145 X)

Photographs of students - Tony Langham

Layout - Peter Faldon

Graphics - Vanda Kiernan

Typing - Stella Jefferies

CONTENTS

Experiments in inorganic chemistry

Experiments in organic chemistry

INTERNATIONAL HAZARD SYMBOLS

 Harmful

 Flammable

 Corrosive

 Toxic

 Explosive

 Oxidising

 Radioactive

INTRODUCTION

The original ILPAC Units, from which these experiments are taken, are organised into four Blocks (Starter, Physical, Inorganic and Organic) but, since the Starter Block Units are concerned mainly with introductory physical chemistry, the experiments are here arranged in three sections, as follows:

EXPERIMENTS 1-49   Physical Chemistry   (115 pages)

EXPERIMENTS 50-80   Inorganic Chemistry   (80 pages)

EXPERIMENTS 81-99   Organic chemistry   (58 pages)

The numbers of experiments in each section is not an indication of relative importance or time alllocation as experiments vary considerably in their length and complexity.

Many of the experiments may be used either for Practical Assessments or as preparation for Practical Examinations. In particular, Experiments 10, 28, 35 and 43 are suitable for assessment of planning skills, since we do not give the usual requirements lists or detailed instructions, while Experiments 60, 61, 67, 75, 80, 88 and 95 are especially useful in preparing for practical examinations which include the investigation of unknown substances.

The order in which the experiments are listed is almost identical to that in the ILPAC Units, and could provide a logical teaching sequence within each section. However, it is envisaged that the material in Experiments 1-16 might be covered first (they come from the Starter Block Units). Thereafter, experiments in the three sections can be "dovetailed".

Advanced Practical Chemistry Resource Pack

This pack of loose-leaf sheets is available as support material. It contains the following items for each experiment.

TECHNICIANS' SHEET   -   a detailed requirements lists, with notes on apparatus and instructions for making up solutions, etc.

TEACHERS' NOTES   -   brief comments, including likely sources of error and difficulty, possible alternatives, and the accuracy to be expected.

*SPECIMEN RESULTS   -   including step-by-step calculations, where appropriate.

*ANSWERS   -   suggested answers to the 'consolidation' questions.

*BLANK TABLES   -   for students to fill in (these are provided only for the observation and deduction experiments and for a few others where the Results Tables are large and/or complex).

*For these pages only, multiple copies may be produced for use within the establishment for which the pack was purchased.

EXPERIMENT 1
Determining the Avogadro constant

Aim

The purpose of this experiment is to estimate the value of the Avogadro constant and to compare this estimate with the accepted value.

Introduction

When a solution of oleic acid, $C_{17}H_{33}CO_2H$, in pentane is dropped on to water, the pentane evaporates leaving behind a layer of oleic acid one molecule thick. For this reason, this experiment has been called 'The Monomolecular Layer Experiment'.

You use a loop of hair or thread to contain the oleic acid and to give a measure of the surface area. By making certain assumptions about the shape of the molecule and its alignment on the surface, you can get a reasonably accurate value for the Avogadro constant.

The experiment has two parts. In the first, you calibrate the pipette. This gives the volume of one drop of solution. In the second part you determine how many drops of solution are required to just fill the loop with a layer of oleic acid molecules. Then we lead you, step by step, through the calculation.

Requirements

measuring cylinder, 10 cm³
teat pipette and adaptor (for small drops)
trough
human hair or cotton thread, 40 - 50 cm
scissors
petroleum jelly or Vaseline
oleic acid solution in pentane (0.05 cm³ of oleic acid per dm³)-------

---

Hazard Warning

Pentane is highly flammable.

Therefore you MUST:

KEEP THE STOPPER ON THE BOTTLE WHEN NOT IN USE:

KEEP THE LIQUID AWAY FROM FLAMES.

Procedure

1. Fill the teat pipette with oleic acid solution and deliver it drop by drop into the 10 cm³ measuring cylinder. Count the number of drops which must be delivered from the pipette to reach the 1 cm³ mark. Enter your value in a copy of Results Table 1.

Results Table 1

Number of drops to deliver 1 cm³ of solution	Number of drops delivered to make monomolecular layer	Diameter of monomolecular layer/cm

2. Tie the hair or cotton thread in a loop. Use a reef-knot (Fig. 1), rather than an overhand knot, so that the loop will make a flat circle. Gut the ends as close to the knot as possible. Hair is preferred because it does not need greasing but if you are using thread, thoroughly but lightly grease it with petroleum jelly. It is most important that no part of the thread escapes greasing. Run the knotted thread through your fingers several times before wiping off the excess.

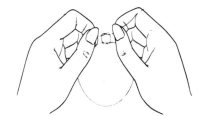

Fig. 1.

3. Fill the trough with water and float the loop on it, making sure that the entire circumference is in contact with the surface. Look very carefully for 'bridges' or submerged loops and move them into the surface with a clean glass rod or a pencil point.

4. Using the same pipette, add the oleic acid solution dropwise to the middle of the loop until it is filled. At first you will probably see the loop expand to a circle and then retract again.
Before the loop is filled, it 'gives' when you push it gently from the outside with a pencil. (Fig. 2)
When the loop is filled, it will slide across the surface, only denting very slightly when pushed gently with the pencil. (Fig. 3)

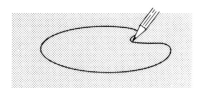

Fig.2.

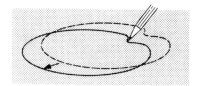

Fig.3.

Count the number of drops required to fill the loop and record this in a copy of Results Table 1.

5. Measure the diameter of the loop and complete Results Table 1.

6. If you have time, repeat the whole procedure. However, you must use a
    fresh hair or thread, and wash out the trough thoroughly to obtain a
    clean surface.

Calculation

1. Calculate the volume of 1 drop delivered from the teat pipette using the
    value in column one of Results Table 1.

    Volume of 1 drop = _____ cm³

2. Calculate the volume of oleic acid in 1 drop of solution delivered from
    the teat pipette.

    Remember that 1000 cm³ of this solution contains 0.05 cm³ of oleic acid.

    Volume of oleic acid in 1 drop = _____ cm³

3. Calculate the volume of oleic acid delivered to make the monomolecular
    layer; i.e. the volume of oleic acid in 1 drop x the number of drops
    required.

    Volume of oleic acid in monolayer = _____ cm³

4. Calculate the surface area of the oleic acid layer.

    Area = $\pi d^2/4$ = _____ cm²

5. You know the volume of oleic acid (from 3) and the surface area it covers
    (from 4). It is a simple matter to calculate the thickness of the layer
    because volume = area x thickness.

Fig.4.

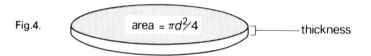

area = $\pi d^2/4$ —— thickness

    Thickness = _____ cm

6. Calculate the volume of one molecule of oleic acid by assuming it is a
    cube, with sides equal to the thickness of the layer.

    Volume of one molecule = _____ cm³

7. Calculate the molar volume of oleic acid given that its density is
    0.890 g cm⁻³ and its molar mass is 282 g mol⁻¹.

    Molar volume of oleic acid = _____ cm³ mol⁻¹

8. Divide the molar volume by the volume of one molecule to determine the
    Avogadro constant.

    L = _____ mol⁻¹

Questions

1. Suggest some sources of error in this experiment which account for the
    discrepancy between the value of L you obtained and the accepted value
    of $L = 6.02 \times 10^{23}$ mol⁻¹.

2. Which of the values you used in your calculations is subject to the
    greatest error?

3. Pentane is not the only liquid that can be used in this experiment.
    Suggest four properties which a suitable substitute must have.

Aim

The purpose of this experiment is to prepare a standard solution of potassium hydrogenphthalate.

Introduction

Potassium hydrogenphthalate, $C_8H_5O_4K$, is a primary standard because it meets certain requirements.

1. It must be available in a highly pure state.

2. It must be stable in air.

3. It must be easily soluble in water.

4. It should have a high molar mass.

5. In solution, when used in volumetric analysis, it must undergo complete and rapid reaction.

You weigh accurately a sample of potassium hydrogenphthalate and use it to make a solution of concentration close to 0.10 mol dm^{-3}. In Experiment 3 you use this solution to determine the concentration of a solution of sodium hydroxide.

Requirements

safety spectacles
weighing bottle
spatula
potassium hydrogenphthalate, $C_8H_5O_4K$
access to a balance capable of weighing to within 0.01 g
beaker, 250 cm^3
wash bottle of distilled water
stirring rod with rubber end
volumetric flask, 250 cm^3, with label
filter funnel
dropping pipette

Procedure

1. Transfer between 4.8 and 5.4 g of potassium hydrogenphthalate into a weighing bottle and weigh it to the nearest 0.01 g.

2. Put about 50 cm^3 of water into a 250 cm^3 beaker. Carefully transfer the bulk of the potassium hydrogenphthalate from the weighing bottle into the beaker.

3. Reweigh the bottle with any remaining potassium hydrogenphthalate to the nearest 0.01 g.

4. Stir to dissolve the solid, adding more water if necessary.

5. Transfer the solution to the volumetric flask through the filter funnel.
    Rinse the beaker well, making sure all liquid goes into the volumetric
    flask. (Some workers transfer the solid directly into the flask through
    a filter funnel, but you should only do this if you are sure the solid
    will dissolve easily and if your funnel has a wide enough stem to prevent
    blockage.)

6. Add distilled water until the level is within about 1 cm of the mark on
    the neck of the flask. Insert the stopper and shake to mix the contents.

7. Using the dropping pipette, add
    enough water to bring the bottom
    of the meniscus to the mark, as in
    Fig. 5. Insert the stopper and
    shake thoroughly ten times to
    ensure complete mixing. Simply
    inverting the flask once or twice
    does not mix the contents properly
    and is a very common fault.

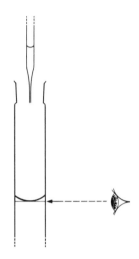

Fig.5.

8. Label the flask with the contents, your name and the date. Leave a space
    for the concentration to be filled in after you have calculated it. Set
    aside the flask for Experiment 3.

Results and Calculations

Using your data, you can fill in a copy of Results Table 2.

Results Table 2

Molar mass of potassium hydrogenphthalate, M	$g\ mol^{-1}$
Mass of bottle and contents before transfer, m_1	g
Mass of bottle and contents after transfer, m_2	g
Mass of potassium hydrogenphthalate, $m = (m_1 - m_2)$	g
Amount of potassium hydrogenphthalate, $n = m/M$	mol
Volume of solution, V	dm^3
Concentration of potassium hydrogenphthalate, $c = n/V$	$mol\ dm^{-3}$

Questions

1. What effect would each of the errors described below have on the concen-
    tration of potassium hydrogenphthalate?

    (a) Some of the solid potassium hydrogenphthalate was spilled in making
        the transfer.

    (b) Not enough water was added to bring the volume up to the mark.

6

EXPERIMENT 3
An acid-base titration

Aim

The purpose of this experiment is to determine the concentration of a solution of sodium hydroxide by titration against a standard solution of potassium hydrogenphthalate.

Introduction

In Experiment 2 you made a standard solution of potassium hydrogenphthalate, a primary standard. The substance has the formula $C_8H_5O_4K$, but because it behaves as a monoprotic (monobasic) acid in producing one mole of hydrogen ions per mole of compound, we can simplify the formula to HA. This simple formula is often used to represent an acid with a complicated structure.

Sodium hydroxide reacts with potassium hydrogenphthalate according to the equation:

$$\text{(aq)} + Na^+OH^-\text{(aq)} \rightarrow \text{(aq)} + H_2O\text{(l)}$$

$$\text{or} \qquad HA\text{(aq)} + Na^+OH^-\text{(aq)} \rightarrow Na^+A^-\text{(aq)} + H_2O\text{(l)}$$

To show you when the reaction is complete - the stoichiometric point or equivalence point - you use an indicator called phenolphthalein, which is colourless in acid and pink in alkaline solution. The point at which the addition of one drop (or even less) of alkali changes the solution from colourless to just faintly pink is called the end-point and, in this case, shows that the reaction is just complete.

Requirements

safety spectacles
filter funnel, small
burette, 50 cm^3, and stand
2 beakers, 100 cm^3
sodium hydroxide solution, approx. 0.1 M NaOH $- - - - - - - - - - - -$
pipette, 25 cm^3
pipette filler
standard potassium hydrogenphthalate solution (prepared in Experiment 2)
4 conical flasks, 250 cm^3
phenolphthalein indicator solution
white tile
wash-bottle of distilled water

Procedure

1. Using the funnel, rinse the burette with the sodium hydroxide solution and fill it with the same solution. Do not forget to rinse and fill the tip. Record the initial burette reading in the 'Trial' column of Results Table 3.

2. Using a pipette filler, rinse the pipette with some of the potassium hydrogenphthalate solution and carefully transfer 25.0 cm³ of the solution to a clean 250 cm³ conical flask.

3. Add 2-3 drops of the phenolphthalein indicator solution.

4. Run sodium hydroxide solution from the burette into the flask, with swirling, until the solution just turns pink. This first flask may be used as a trial run, because you will probably overshoot the end-point. Record the final burette reading.

5. Refill the burette with the sodium hydroxide solution, and again record the initial burette reading to the nearest 0.05 cm³ (one drop).

6. Using the pipette, transfer 25.0 cm³ of the potassium hydrogenphthalate solution to another clean conical flask. Add 2-3 drops of the phenolphthalein indicator solution.

7. Carefully titrate this solution to the end-point, adding the alkali drop by drop when you think the colour is about to change.

8. Repeat steps 5, 6 and 7 at least twice more.

9. Empty the burette and wash it carefully immediately after the titration, especially if it has a ground glass tap.

Accuracy

You should record burette readings to the nearest 0.05 cm³ (approximately one drop). Consecutive titrations should agree to within 0.10 cm³ and, strictly, you should repeat the titration until this is achieved. However, you may have neither the time nor the materials to do this. With practice, your technique will improve so that it is not necessary to do more than four titrations. Calculate the mean of the two (or preferably three) closest consecutive readings and quote this also to the nearest 0.05 cm³. Note that this does not introduce a fourth significant figure; it merely makes the third figure more reliable.

Results Table 3

Pipette solution					mol dm⁻³	cm³
Burette solution					mol dm⁻³	
Indicator						
		Trial	1	2	3	(4)
Burette readings	Final					
	Initial					
Volume used (titre)/cm³						
Mean titre/cm³						

Calculation

1. Calculate the concentration of the sodium hydroxide solution.

Questions

1. What effect would each of the errors described below have on the calculated value of the concentration of sodium hydroxide?

   (a) The burette is not rinsed with the sodium hydroxide solution.

   (b) The pipette is not rinsed with the potassium hydrogenphthalate solution.

   (c) The tip of the burette is not filled before titration begins.

   (d) The conical flask contains some distilled water before the addition of potassium hydrogenphthalate.

2. In using phenolphthalein as an indicator, we prefer to titrate from a colourless to pink solution rather than from pink to colourless. Suggest a reason for this.

3. Why is it advisable to remove sodium hydroxide from the burette as soon as possible after the titration?

EXPERIMENT 4
A redox titration

Aim

The purpose of this experiment is to balance the
equation for the reaction between sodium
thiosulphate and iodine.

$$a\ Na_2S_2O_3(aq) + b\ I_2(aq) \rightarrow \text{Products}$$

Introduction

You are to determine the ratio of a to b and so determine the stoichiometry
of the reaction. You do this by taking a known amount of iodine and titra-
ting it with standard sodium thiosulphate.

The indicator you use in this titration is starch solution, which is deep blue
in the presence of iodine; it is added near the end of the titration when the
solution is straw-coloured. If you add starch too soon, you may get a blue-
black precipitate which does not dissolve again easily even though there is an
excess of thiosulphate. The end-point in this titration is the point at which
the addition of one drop of sodium thiosulphate causes the disappearance of
the deep blue colour.

Requirements

safety spectacles
filter funnel
burette, 50 cm^3, and stand
2 beakers, 100 cm^3
sodium thiosulphate solution, standardized
pipette, 10 cm^3
pipette filler
iodine solution, standardized
4 conical flasks, 250 cm^3
starch indicator solution
white tile
wash-bottle of distilled water

Procedure

1. Using the funnel, rinse the burette and tip with the sodium thiosulphate
    solution. Fill it with the same solution. Don't forget to fill the tip.
    Record the initial burette reading in Results Table 4.

2. Rinse the pipette with some of the iodine solution and carefully transfer
    10.0 cm^3 of the solution to one of the conical flasks.

3. Titrate this solution until the colour of the iodine has <u>almost</u> gone (as
    indicated by a pale straw colour).

4. Add 1-2 cm³ of starch solution and continue the titration, adding sodium thiosulphate dropwise until the end-point. Use the first flask for a trial run. Record the final burette reading.

5. Repeat the titration three more times. Enter your results into a copy of Results Table 4. These titrations should agree to within 0.10 cm³.

Results Table 4

Pipette solution					mol dm⁻³	cm³
Burette solution					mol dm⁻³	
Indicator						
		Trial	1	2	3	(4)
Burette readings	Final					
	Initial					
Volume used (titre)/cm³						
Mean titre/cm³						

Calculation

1. Use your results to determine the stoichiometric coefficients, a and b, in the equation:

$$a \ Na_2S_2O_3(aq) + b \ I_2(aq) \rightarrow Products$$

2. All the iodine forms sodium iodide NaI. There is one other product - work out its formula.

EXPERIMENT 5
A precipitation titration

Aim

The purpose of this experiment is to determine the number of molecules of water of hydration in hydrated barium chloride, i.e. to calculate the value of x in the formula $BaCl_2 \cdot xH_2O$

Introduction

You titrate chloride ions with silver ions, according to the equation:

$$Ag^+(aq) + Cl^-(aq) \rightarrow AgCl(s)$$

This provides you with the data necessary to do the calculations. The indicator for the titration is potassium chromate(VI). When all the chloride ions have reacted, any more silver ions react with the indicator producing a red precipitate of silver chromate(VI). This is because silver chloride is less soluble than silver chromate(VI).

$$2Ag^+(aq) + CrO_4{}^{2-}(aq) \rightarrow Ag_2CrO_4(s)$$

The end-point in this reaction is when one drop of aqueous silver ions produces a red tinge on the precipitate of silver chloride.

Barium ions also react with chromate ions so the barium must be removed by adding sulphate ions:

$$Ba^{2+}(aq) + SO_4{}^{2-}(aq) \rightarrow BaSO_4(s)$$

This does not affect the concentration of chloride ions.

Requirements

safety spectacles
weighing bottle
spatula
barium chloride crystals —————————————————————
access to balance capable of weighing to 0.01 g
beaker, 250 cm^3
wash-bottle of distilled water
stirring rod with rubber end
volumetric flask, 250 cm^3, with label
filter funnel
dropping pipette
burette, 50 cm^3, and stand
2 beakers, 100 cm^3
silver nitrate solution, standardized ————————————
pipette, 10 cm^3
pipette filler
4 conical flasks, 250 cm^3
sodium sulphate
potassium chromate solution
'silver residues' bottle

Hazard Warning

Barium chloride is very poisonous.
Silver nitrate is also poisonous and
can stain the skin.
Therefore you MUST:

USE THE PIPETTE FILLER SUPPLIED

WASH YOUR HANDS AFTER USE

Procedure

1. Prepare a standard solution of hydrated barium chloride by accurately
    weighing out between 1.4 g and 1.6 g of the salt. Dissolve this and make
    up to 250 cm^3 in a volumetric flask. Fill in a copy of Results Table 5a.

2. Rinse the burette with some silver nitrate solution and fill. Don't
    forget the tip.

3. Rinse the 10.0 cm^3 pipette with barium chloride solution, and transfer
    10.0 cm^3 to a conical flask.

4. Add about 1 g of sodium sulphate crystals to the flask and swirl it.

5. Add 2-3 drops of potassium chromate(VI) indicator. Titrate the solution
    to the end-point, as shown by the first appearance of a permanent but
    faint reddish precipitate of silver chromate(VI). Use the first flask
    for a trial run. Enter your results in a copy of Results Table 5b.

6. Repeat steps 2 to 5 three times. Don't wash the contents of the titration
    flasks down the sink - pour them into a 'silver residues' bottle.

Results and calculations

Results Table 5a

Mass of bottle and contents before transfer, m_1		g
Mass of bottle and contents after transfer, m_2		g
Mass of sample, $m = (m_1 - m_2)$		g
Mass of $BaCl_2 \cdot xH_2O$ in 10.0 cm³		g

Results Table 5b

Pipette solution					mol dm⁻³	cm³
Burette solution					mol dm⁻³	
Indicator						
		Trial	1	2	3	(4)
Burette readings	Final					
	Initial					
Volume used (titre)/cm³						
Mean titre/cm³						

Calculation

1. From the mean titre and concentration of silver nitrate, calculate the amount of chloride ion present in a 10.0 cm³ sample.

2. Calculate the mass of anhydrous barium chloride, $BaCl_2$, present in a sample.

3. Calculate the mass of water present by subtracting the mass of $BaCl_2$ from the mass of $BaCl_2 \cdot xH_2O$.

4. Determine the ratio of amount of $BaCl_2$ to amount of H_2O and thus the value of x.

A flow-chart for this multi-step calculation is as follows:

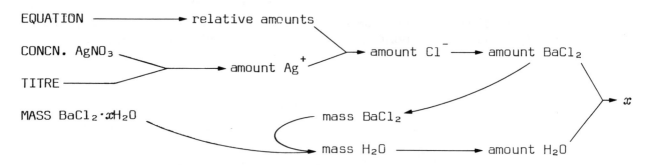

14

Experiment 6
A titration exercise

Aim

The purpose of this experiment is to determine x in the formula $Fe(NH_4)_2(SO_4)_2 \cdot xH_2O$ by titration against a standard solution of potassium manganate (VII) (permanganate).

Introduction

This experiment provides useful experience in preparation for a practical examination. The directions are similar to those that would be given by an examination board.

   A   is a solution of ammonium iron(II) sulphate, $Fe(NH_4)_2(SO_4)_2 \cdot xH_2O$, the precise concentration of which (in g dm^{-3}) is given by the teacher.

   B   is a solution of potassium manganate(VII) (permanganate), $KMnO_4$, the precise concentration of which (in mol dm^{-3}) is given by the teacher.

Requirements

safety spectacles
pipette, 25 cm^3
pipette filler
3 conical flasks, 250 cm^3
ammonium iron(II) sulphate solution, A
sulphuric acid, dilute, 1 M H_2SO_4
burette, 50 cm^3
burette stand
funnel, small
potassium manganate(VII) solution, B
white tile
wash-bottle of distilled water

Procedure

Pipette 25 cm^3 of the ammonium iron(II) sulphate solution, A, into a conical flask and add an equal volume of dilute sulphuric acid. Titrate with potassium manganate(VII) solution, B, until a permanent faint pink colour appears. Repeat the titration twice and enter your results in a copy of Results Table 6.

The overall equation for the reaction is

$$MnO_4^-(aq) + 5Fe^{2+}(aq) + 8H^+(aq) \rightarrow Mn^{2+}(aq) + 5Fe^{3+}(aq) + 4H_2O(l)$$

Results Table 6

Pipette solution					g dm^{-3}	cm^3
Burette solution					mol dm^{-3}	
Indicator						
		Trial	1	2	3	(4)
Burette readings	Final					
	Initial					
Volume used (titre)/cm^3						
Mean titre/cm^3						

Calculation

Use your results to determine x in the formula $Fe(NH_4)_2(SO_4)_2 \cdot xH_2O$. You should set out your calculations so that every step in your working is clearly shown. If you cannot work out a method of calculation, use the suggestions below. (These would probably not be given in an examination.)

Calculation steps

1. From the titre and the equation for the reaction calculate the concentration of Fe^{2+} ions (y mol dm^3).

2. From the concentration calculate the mass of anhydrous $Fe(NH_4)_2(SO_4)_2$ in one litre of solution.

3. Subtract the mass obtained in step 2 from the mass of the salt in one litre. This difference, z g, divided by 18 g mol^{-1} gives the amount of water of crystallization in y mol of the salt.

4. x mol is the amount of water in 1 mol of the salt, i.e.

$$x = \frac{z}{18} \times \frac{1}{y}$$

16

EXPERIMENT 7
Estimating the ionization energy
of a noble gas

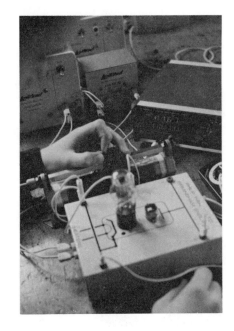

Aim

The purpose of this experiment is to
estimate the ionization energy of
argon by the electron impact method.

Introduction

Argon atoms can be ionized by bombarding them with high-energy electrons:

$$Ar(g) \quad + \quad e^- \quad \rightarrow \quad Ar^+(g) \quad + \quad e^- \quad + \quad e^-$$

↑	↑	↑
high-energy electron	electron from Ar	bombarding electron with less energy

The process is carried out in an argon-filled radio valve, called a thyratron,
as shown in Fig. 6.

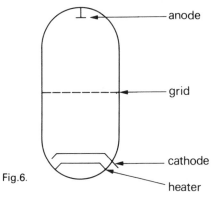

The valve anode is kept at a negative potential
to serve as a collector of positive ions. Note
that the term 'anode' is not really appropriate
here as an anode is normally positively charged.

The grid is kept positive to attract electrons
emitted from the heated cathode.

The heater provides energy which allows the
cathode to give off electrons.

Fig.6.

The experiment takes place in several stages which we now illustrate with
diagrams of the valve:

1. The cathode is heated (by passing a current
   through the heater circuit) and emits electrons.
   These are attracted towards the positively
   charged grid, increasing their speed and energy.
   Some electrons are collected by the grid, while
   some pass through holes and move towards the
   anode. Since the anode is negatively charged
   relative to the grid, it repels electrons, which
   turn back from it and return to the grid.

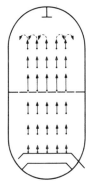

Fig.7.

2. As the potential between the grid and the cathode
   is increased by adjusting a variable resistor, the
   electrons passing through the grid increase in
   energy but still cannot reach the anode. Note
   that the lengths of the straight arrows in Figs.
   7 and 8 indicate their relative speeds.

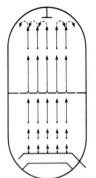

Fig.8.

3. When the electrons have enough energy, they
   knock other electrons off the argon atoms,
   forming positive argon ions. These argon ions
   are then attracted to the negative anode
   where they pick up electrons from the circuit and cause a current to flow. This is
   registered by a milliammeter and marks the
   potential at which ionization takes place.

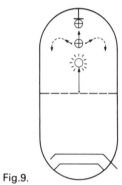

key:
·   electron
○   argon atom
⊕   argon ion

Fig.9.

   In brief then, the potential between the grid and the cathode is slowly
   increased until there is a rapid jump in the current shown by the milliammeter. This is the potential at which ionization takes place and we can
   calculate the ionization energy from it.

Requirements

circuit board with labelled sockets for valve and connecting leads
argon-filled valve, type 884
power supplies
   (a) ·6.3 V (AC or DC) for the cathode heater
   (b) 3 V (DC) for the anode circuit
   (c) 25 V (DC) for the grid circuit
potential divider, 250 Ω ⎫
                          ⎬ these may be included in the circuit board
protective resistor, 200 Ω ⎭
voltmeter, 25 V, high resistance
microammeter, 100 μA (0.1 mA)
12 connecting leads

Procedure

1. Insert the valve and connect the
    6.3 V supply to the sockets label-
    led 'cathode heater'. Switch on
    to allow the valve to warm up for
    a minute or two.

2. Connect the voltmeter, microammeter
    and 3 V supply to the appropriate
    labelled sockets. Make sure the
    valve anode is negative with res-
    pect to the cathode, as shown in
    Fig. 10. (In most valve appli-
    cations, the anode is positive,
    as the name implies.)

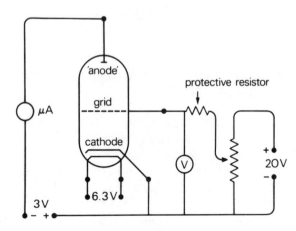

Fig.10.

3. Set the potential divider to give the minimum accelerating potential; connect the 25 V supply and the protective resistor, as shown in Fig. 10. If you are not sure about these connections, check with your teacher.

4. Slowly increase the potential and watch the voltmeter. At the same time, watch the microammeter, which should read zero until ionization occurs.

5. As soon as the microammeter needle begins to move, note the voltmeter reading and reduce the potential to zero.

6. Repeat steps 4 and 5, this time looking for two additional indications of ionization:

   (i) a sudden decrease in the voltmeter reading (due to the surge in the grid current),

   (ii) a small purple glow in the valve (looking down from the top).

You may need to move the potential divider control a little further to see these effects, but always return it to zero as soon as possible to avoid damage to the valve.)

Calculation

One mole of electrons, on moving through a potential difference of one volt, acquires 96.3 kJ of energy*. To calculate the ionization energy of argon, multiply your voltage by this factor.

$$\text{1st ionization energy of argon} = \underline{\hphantom{xxx}} \text{ V} \times 96.3 \text{ kJ mol}^{-1} \text{ V}^{-1}$$

$$= \underline{\hphantom{xxx}} \text{ kJ mol}^{-1}$$

Note that if you compare the value you have just calculated with the data-book value, you will probably find it is up to 20% higher. This can be explained by the limitations of the experimental method. The experiment is designed to give you an idea of the meaning of ionization energy, rather than to provide an accurate method of determining it. You will probably consider the more accurate method, based on emission spectra, as a theoretical topic.

* This conversion factor is derived as follows:

It follows from basic definitions that one coulomb (1 C) of charge accelerated by a potential difference of one volt (1 V) gains one joule (1 J) of energy, i.e. 1 C accelerated by 1 V gains 1 J.

The charge on an electron is 1.6×10^{-19} C

$\therefore$ 1 electron accelerated by 1 volt gains 1.6×10^{-19} J

There are 6.02×10^{23} electrons in 1 mol of electrons

$\therefore$ 1 mol of electrons accelerated by 1 V gains $1.6 \times 10^{-19} \times 6.02 \times 10^{23}$ J

$$= 96300 \text{ J}$$

The energy acquired by a single electron accelerated by one volt is sometimes referred to as one electron-volt (ev), and ionization energies are sometimes quoted in electron-volts. The calculation above shows that:

$$\boxed{1 \text{ eV} \equiv 96.3 \text{ kJ mol}^{-1}}$$

Question

1. If another microammeter were included in the circuit for this experiment, between the grid and the protective resistor, it would show a current flowing throughout the experiment, rising suddenly after ionization.

   (a) Explain why a current flows before ionization.

   (b) Suggest one reason for the sudden increase after ionization.

EXPERIMENT 8
Using a hand spectroscope to observe the
emission spectra of some *s*-block elements

Aim

This experiment is designed to give you a
qualitative introduction to the spectra emitted
by some *s*-block elements when their atoms are
excited by heating samples in a Bunsen flame.

Introduction

You use a hand spectroscope to observe the continuous spectrum emitted by the
tungsten filament of a light bulb. Using a flame-test wire, you then obtain
coloured emissions from some *s*-block elements, view these in turn through the
spectroscope and compare them with the continuous spectrum from the tungsten
filament.

We recommend that you work in pairs in this experiment. All three operations
- preparing the flame-test wire, obtaining a brightly coloured flame and
having the spectroscope ready to look at the flame for the few moments that
it lasts - are fairly tricky and require practice and concentration. We
suggest that one person prepares the flame while the other stands ready with
the spectroscope. You can then change roles so that you both have a chance
to observe each spectrum.

Requirements

fume cupboard with gentle fan
electric lamp with tungsten filament pearl light bulb
hand spectroscope
colour plate of spectra
Bunsen burner and bench protection sheet
flame-test wire
hydrochloric acid, HCl, concentrated —
boiling tube or other glass container for hydrochloric acid
small pestle and mortar
spatulas
watch-glasses
at least three of the following:
   barium chloride, $BaCl_2(s)$
   calcium chloride, $CaCl_2(s)$
   lithium chloride, $LiCl(s)$
   potassium chloride, $KCl(s)$
   sodium chloride, $NaCl(s)$
   strontium chloride, $SrCl_2(s)$
safety spectacles

Procedure

1. Switch on the lamp and look at the bulb through the spectroscope. Look
    for a series of colours, one running into the next. This is a continuous
    spectrum. Compare what you see with a coloured plate showing the
    emission spectrum from a tungsten filament.

2. Hold the spectroscope up to a window which does not face the sun.
    You must NEVER point the spectroscope directly at the sun. This
    could result in permanent damage to your eyes.

    You should see the continuous spectrum of visible light.

3. Light the Bunsen burner - adjust it to get a roaring flame.

4. Dip the flame test wire into concentrated hydrochloric acid, then
    hold it in the hottest part of the flame. Repeat the process
    until there is little or no colour from the flame test wire in
    the flame. You may have to repeat this step several times,
    especially towards the end of the experiment, but certainly no more than
    twelve times.

5. Crush a little of the salt to be tested finely in a pestle and mortar and
    mix with a little concentrated hydrochloric acid on a watch glass. Be
    careful here - use just enough of the acid to give you a semi-solid
    'mush' of crystals.

6. Dip the cleaned flame-test wire into the mush of the salt to be tested.
    Adjust the flame until it is pale blue and hold the wire in it. Your
    partner should be standing by with the spectroscope and should now look
    through it at the flame. Look for brightly coloured lines.

    There are several lines for each element and it will probably not be
    possible to get them all into view at once. The yellow line in the sodium
    spectrum is easy to see and will probably persist through the spectra of
    all the elements you try. You can use this line to help you locate lines
    on the spectra of the other elements, by looking either to the right or
    left of it. Checking with the coloured plate of spectra will give you an
    idea of where to look for lines from a particular element.

7. Repeat steps 2-6 with the salts of at least two other elements. Also use
    the spectroscope on any other vapour lamps which may be available,
    including street lamps, if there is one in view from the laboratory.

Question

What is the difference between a continuous spectrum and a line emission
spectrum?

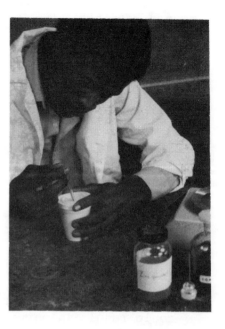

EXPERIMENT 9
Determining an enthalpy change of reaction

Aim

The purpose of this experiment is to determine the enthalpy change for the displacement reaction:

$$Zn(s) + Cu^{2+}(aq) \rightarrow Cu(s) + Zn^{2+}(aq)$$

Introduction

By adding an excess of zinc powder to a measured amount of aqueous copper(II) sulphate, and measuring the temperature change over a period of time, you can then calculate the enthalpy change for the reaction.

Requirements

safety spectacles
pipette, 25 cm³
pipette filler
polystyrene cup with lid
copper(II) sulphate solution, 1.00 M CuSO₄ -
weighing bottle
spatula
zinc powder
balance
thermometer, 0 - 100 °C (0.2 °C graduations)
watch or clock with second hand

Procedure

1. Pipette 25.0 cm³ of the copper(II) sulphate solution into a
   polystyrene cup.

2. Weigh about 6 g of zinc powder in the weighing bottle. Since this is an excess, there is no need to be accurate.

3. Put the thermometer through the hole in the lid, stir and record the temperature to the nearest 0.1 °C every half minute for $2\frac{1}{2}$ minutes.

4. At precisely 3 minutes, add the zinc powder to the cup.

5. Continue stirring and record the temperature for an additional 6 minutes to complete Results Table 9.

Results Table 9

Time/min	0.0	0.5	1.0	1.5	2.0	2.5	3.0	3.5	4.0	4.5
Temperature/°C							–			
Time/min	5.0	5.5	6.0	6.5	7.0	7.5	8.0	8.5	9.0	9.5
Temperature/°C										

Calculations

1. Plot the temperature (y-axis) against time (x-axis).

2. Extrapolate the curve to 3.0 minutes to establish the maximum temperature rise as shown in Fig.11.

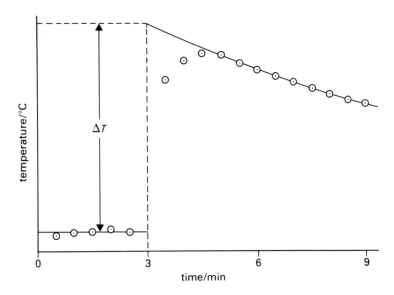

Fig.11.

3. Calculate the enthalpy change for the quantities used, making the same assumptions as in the preceding exercise.

4. Calculate the enthalpy change for one mole of Zn and $CuSO_4$(aq), and write the thermochemical equation for the reaction.

Questions

1. Compare your result with the accepted value of -217 kJ mol^{-1} by calculating the percentage error in your answer:

$$error = \frac{experimental\ value\ -\ accepted\ value}{accepted\ value} \times 100\%$$

2. List some possible reasons for any difference between your value and the accepted value.

3. Why do you think the temperature increases for a few readings after adding the zinc? (Hint: it does not increase if more zinc is used or if the powder is very finely divided.)

EXPERIMENT 10
Determining an enthalpy change of solution

Aim

The purpose of this experiment is to determine
the enthalpy change for the process

$$NH_4Cl(s) + 100H_2O(l) \rightarrow NH_4Cl(aq, 100H_2O)$$

Introduction

Because this is a planning experiment, we give fewer details and instructions
than you have been used to. It is, of course, very similar to Experiment 9,
but you need not plot a temperature/time graph because the maximum temperature
change occurs very rapidly.

Requirements

Make a list of requirements including the masses and amounts needed; show the
list to your teacher or technician.

Procedure

Work this out for yourself and keep an accurate record.

Results

1. Tabulate your results in an appropriate form.

2. Calculate the enthalpy change of solution for the thermochemical equation
   in the aim.

Question

Compare your result with the accepted value of $+16.4$ kJ mol^{-1}. Suggest
reasons for any difference.

EXPERIMENT 11
Using Hess's law

(Experiment 12 is an alternative,
using a more sophisticated method.)

Aim

The purpose of this experiment is to determine
the enthalpy change for the reaction

$$MgSO_4(s) + 7H_2O(l) \rightarrow MgSO_4 \cdot 7H_2O(s)$$

Introduction

It is impossible to measure the enthalpy change for this reaction directly
because the process cannot be controlled. However, you can calculate this
enthalpy change by measuring the enthalpy change of solution for the two
solids:

$$MgSO_4(s) + 100H_2O(l) \rightarrow MgSO_4(aq,100H_2O)$$

$$MgSO_4 \cdot 7H_2O(s) + 93H_2O(l) \rightarrow MgSO_4(aq,100H_2O)$$

We suggest that you use 0.0250 mol of each salt, so we have calculated, from
the equations, the required masses of each salt and water.

Requirements

safety spectacles
2 weighing bottles
spatula
magnesium sulphate (anhydrous), $MgSO_4$
access to balance
2 polystyrene cups and lids
distilled water
teat pipette
thermometer (0° to 50°C)
magnesium sulphate-7-water, $MgSO_4 \cdot 7H_2O$

Procedure

A. Heat of solution of $MgSO_4(s)$

1. Weigh 3.01 g of $MgSO_4$ to the nearest 0.01 g into a clean, dry weighing
    bottle. Record, in a copy of Results Table 11, the masses of weighing
    bottle empty and with contents, unless your balance has a reliable taring
    device.

2. Similarly, weigh 45.00 g of H_2O to the .nearest 0.01 g into a polystyrene
    cup.

3. Put the thermometer through the hole in the lid and measure the temperature
    of the water. Record this in Results Table 11.

4. Carefully transfer the $MgSO_4$ into the water, stir gently with the thermo-
    meter, and record the maximum temperature.

B. Heat of solution of $MgSO_4 \cdot 7H_2O$

5. Weigh 6.16 g of $MgSO_4 \cdot 7H_2O$ to the nearest 0.01 g into a clean, dry weighing bottle.

6. Weigh 41.85 g of H_2O to the nearest 0.01 g into a polystyrene cup.

7. Measure and record the temperature change associated with dissolving the $MgSO_4 \cdot 7H_2O$

Results Table 11

	$MgSO_4$	$MgSO_4 \cdot 7H_2O$
Mass of weighing-bottle		
Mass of weighing-bottle + salt		
Mass of salt	3.01 g	6.16 g
Mass of polystyrene cup		
Mass of polystyrene cup + water		
Mass of water	45.00 g	41.85 g
Initial temperature		
Final temperature		

Calculations

1. From the data in Results Table 11, calculate the enthalpy change of solution for one mole of $MgSO_4$. Assume c_p = 4.18 kJ kg^{-1} K^{-1}.

2. Similarly, calculate the enthalpy change of solution for one mole of $MgSO_4 \cdot 7H_2O$.

3. By means of an energy cycle, calculate the enthalpy change for the reaction:

$$MgSO_4(s) + 7H_2O(l) \rightarrow MgSO_4 \cdot 7H_2O$$

Questions.

1. Plot the results on an energy-level diagram.

2. Why is it not necessary to plot a temperature/time graph as you did in Experiment 9?

3. Compare your result with the accepted value of -104 kJ mol^{-1}. Suggest reasons for any difference.

26

EXPERIMENT 12
Another application of Hess's law

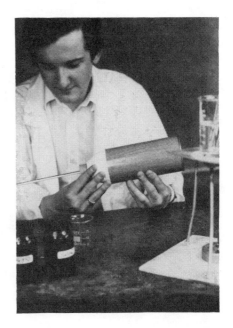

Aim

The purpose of this experiment is to determine
the enthalpy change for the reaction

$$CuSO_4(s) + 5H_2O(l) \rightarrow CuSO_4 \cdot 5H_2O(s)$$

Introduction

Because $CuSO_4(s)$ is slow to dissolve, and $\Delta H^\ominus$ is small, it is best to do this
experiment in a vacuum flask. However, the flask has a measurable heat
capacity which you must determine before you proceed with the experiment.

To calculate the required enthalpy change, you perform two 'heat of solution'
determinations. You should calculate the masses of the salts and water
required. Base your calculations on the following equations:

$$CuSO_4(s) + 100H_2O(l) \rightarrow CuSO_4(aq,100H_2O)$$

$$CuSO_4 \cdot 5H_2O(s) + 95H_2O(l) \rightarrow CuSO_4(aq,100H_2O)$$

and use 0.025 mol of the appropriate salt in each of the determinations.
Show your calculations to your teacher just in case you have made an error
which would spoil your laboratory work.

Requirements

safety spectacles
vacuum flask with thermometer fitted
pipette, 50 cm^3
distilled water
2 beakers, 100 cm^3
Bunsen burner, tripod, gauze and bench mat
thermometer, 0-100 °C
access to balance (weighing to nearest 0.01 g)
2 weighing bottles
spatula
anhydrous copper(II) sulphate, $CuSO_4$
copper(II) sulphate-5-water, $CuSO_4 \cdot 5H_2O$, finely ground beforehand

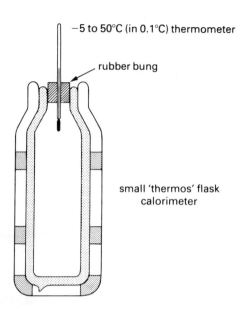

−5 to 50°C (in 0.1°C) thermometer

rubber bung

small 'thermos' flask calorimeter

Fig.12.

Procedure

A. <u>Determination of the heat capacity of the vacuum flask</u>

1. Check that the inside of the vacuum flask is dry and pipette into it 50 cm³ of distilled water at room temperature.

2. Place the thermometer-fitted bung in position and shake gently. Make sure that the entire inside surface of the flask is wet. To read the steady temperature, hold the bung firmly in and turn the flask on its side so that the water covers the mercury reservoir of the thermometer. Read the steady temperature to the nearest 0.1 °C. Leave the bung and thermometer in position.

3. Pipette 50 cm³ of distilled water into a clean, dry 100 cm³ beaker and heat it gently till its temperature reaches about 40 °C - use a small 0-100 °C thermometer. Remove from heat.

4. Use the thermometer from the vacuum flask (with bung still fitted) to gently stir the water in the beaker to ensure its temperature is uniform throughout. Record this temperature to the nearest 0.1 °C.

5. <u>Immediately</u> pour all of the warmed distilled water into the vacuum flask, close with bung and thermometer, shake gently and note the steady temperature to the nearest 0.1 °C. Remember to wet the inside surface and tilt the flask to read the temperature as before.

6. Complete Results Table 12a and then, if you have time, repeat the procedure. It is good practice to do the determination twice and average the results. Furthermore, because this is the first time you have done such an experiment, the second determination should improve your technique.

7. To make best use of laboratory time, we suggest that you complete parts B and C of the experiment before calculating the heat capacity of the flask.

Results Table 12a

Mass of cold water in vacuum flask	g	g
Mass of warm water added	g	g
Initial temperature of flask and cold water	°C	°C
Initial temperature of warm water	°C	°C
Final temperature of flask and mixture	°C	°C

The specific heat capacity of water is 4.18 kJ kg^{-1} K^{-1}.

The density of water is 1.00 g cm^{-3}.

B. Heat of solution of $CuSO_4$(s)

1. Rinse the inside of the vacuum flask with distilled water <u>and drain well.</u>

2. Weigh the calculated quantity of anhydrous copper(II) sulphate ($CuSO_4$), to the nearest 0.01 g, into a clean dry weighing bottle. (<u>Do not keep bottle lids and/or stoppers off longer than is necessary.</u> Why not?)

3. Weigh the appropriate calculated quantity of water, to the nearest 0.1 g, into a dry 100 cm^3 beaker, and then pour this water into the vacuum flask. (If the balance you are using has sufficient overall weighing capacity, you may weigh the water directly into the vacuum flask - why is it better to do this?) Close with bung and thermometer, shake, and note the steady temperature, tilting the flask as before.

4. Remove the bung and thermometer from the vacuum flask and quickly and carefully tip <u>all</u> of the weighed sample of anhydrous copper(II) sulphate into the water. Replace the bung and thermometer, shake to dissolve the salt, and note the temperature once it has become steady. <u>This last step may take up to 15 mins</u> (shaking periodically) because the anhydrous salt is often very slow to dissolve.

5. Complete Results Table 12b and then move on to part C.

Results Table 12b

Mass of anhydrous copper(II) sulphate	g
Mass of water	g
Initial temperature of vacuum flask and water	°C
Maximum temperature of vacuum flask and solution	°C

C. Heat of solution of $CuSO_4 \cdot 5H_2O$(s)

Wash out your apparatus and then repeat the procedure in part B using the hydrated salt ($CuSO_4 \cdot 5H_2O$). In this second determination the salt dissolves quickly in step 4 and the final steady temperature will be obtained within half a minute. Complete Results Table 12c.

Results Table 12c

Mass of copper(II) sulphate-5-water	g
Mass of water	g
Initial temperature of vacuum flask and water	°C
Maximum temperature of vacuum flask and solution	°C

Calculation A (heat capacity of flask)

1. Since the flask is insulated, no heat energy is transferred between system and surroundings (in this case, the flask is part of the system). Therefore you can write:

$$\left[\begin{array}{l}\text{Change in heat}\\\text{energy of flask}\end{array}\right] + \left[\begin{array}{l}\text{Change in heat}\\\text{energy of cold water}\end{array}\right] + \left[\begin{array}{l}\text{Change in heat}\\\text{energy of warm water}\end{array}\right] = 0$$

2. In each case, the change in heat energy = heat capacity x ΔT

   and for the water, heat capacity = mass x specific heat capacity
   $$= \text{mass x } 4.18 \text{ kJ kg}^{-1} \text{ K}^{-1}$$

3. Substitute values from Results Table 12a into these expressions and so obtain a value for the heat capacity, C, of the flask. Remember that ΔT is positive for the flask and the cold water but negative for the hot water, and also that the mass must be in kg.

Calculations B and C (enthalpy changes)

1. Again, there is no heat energy transfer between system and surroundings so that:

$$\left[\begin{array}{l}\text{Change in heat}\\\text{energy of flask}\end{array}\right] + \left[\begin{array}{l}\text{Change in heat}\\\text{energy of contents}\end{array}\right] + \left[\begin{array}{l}\text{Enthalpy change}\\\text{of solution}\end{array}\right] = 0$$

2. Use this expression and values from Results Tables 12b and 12c to obtain the enthalpy changes for dissolving the weighed amounts of the two salts. Ignore the very small heat capacities of the salts, i.e. use the mass and specific heat capacity of the water in your calculations.

3. Scale up to the amounts shown in the equations.

4. Use Hess's law to calculate the required standard enthalpy change.

Questions

1. Suggest a reason why it would be difficult to determine, by direct experiment, $\Delta H^{\ominus}$ for the reaction

   $$CuSO_4(s) + 5H_2O(l) \rightarrow CuSO_4 \cdot 5H_2O(s)$$

2. Why is it good practice to replace the stoppers/lids of chemical bottles as soon as possible?

3. Why, in step 3 of part B of the experiment, would it better to weigh the water directly into the vacuum flask rather than in a beaker?

Aim

The purpose of this experiment is to determine the heats of combustion of a series of similar alcohols from butan-1-ol to octan-1-ol.

Introduction

In this experiment, you burn a measured mass of an alcohol in a spirit lamp and transfer the heat energy released to a calorimeter containing water. From the resulting temperature rise you can calculate the heat of combustion.

In some earlier calorimetric experiments, you assumed that <u>all</u> the heat energy released in a chemical reaction was absorbed by the contents of the calorimeter. You cannot make that assumption in this experiment for two reasons.

1. The heat energy is released in a flame, and although the apparatus is designed to transfer most of the energy to the calorimeter, a significant quantity is lost to the surrounding air.
2. The heat capacity of the calorimeter itself is <u>not</u> so small as to be insignificant compared with the heat capacity of its contents.

You can take account of both these factors by calibrating the apparatus using an alcohol with known heat of combustion.

Requirements

safety spectacles
heat of combustion apparatus
spirit lamp
wood block
retort stand (with 2 clamps and bosses)
Drechsel bottle
filter pump
rubber tubing (for connections to filter pump)
water (at room temperature)
adhesive labels
propan-1-ol, C_3H_7OH ⎫
butan-1-ol, C_4H_9OH ⎪   (If these alcohols are supplied
pentan-1-ol, $C_5H_{11}OH$ ⎬   in separate spirit lamps, you
hexan-1-ol, $C_6H_{13}OH$ ⎪   will not need the next two
heptan-1-ol, $C_7H_{15}OH$ ⎪   items.)
octan-1-ol, $C_8H_{17}OH$ ⎭
6 beakers, 50 cm³
6 teat pipette droppers
Bunsen burner and protective mat
wood splints
tweezers (not plastic-tipped)
balance (preferably capable of weighing to 0.001 g, but 0.01 g will do)
thermometer -5 to -50°C (in 0.1°C)

Procedure

A. Determination of the heat capacity of the apparatus

1. Arrange the heat of combustion apparatus (i.e. the calorimeter), the
   Drechsel bottle, etc., as shown in Fig. 13. Don't use the calorimeter
   base supplied by the manufacturer but stand the burner on a small block
   of wood to ensure a good flow of air. Adjust the height of the
   calorimeter (or the size of the block) so that the top of the spirit
   lamp is level with the bottom of the calorimeter.

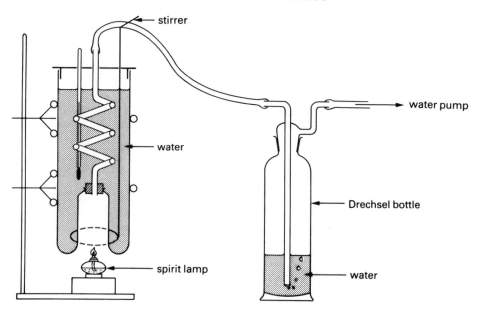

Fig. 13. Heat of combustion apparatus

2. Use water at room temperature (not direct from the tap) to fill the
   calorimeter to about 1.5 cm below the rim. Mark this level with a
   label.

3. Pour a little propan-1-ol into a small beaker and use
   a teat pipette to half-fill the spirit lamp. (This may
   have been done for you already.) Replace the wick and
   cap, return any excess alcohol to the bottle, and remove
   both bottle and beaker to a safe distance from any flame.

4. Turn on the filter pump and adjust the flow of air through the Drechsel
   bottle to about 3-4 bubbles per second.

5. Stand the spirit lamp away from the calorimeter, and use a wood splint
   to light it. Adjust the height of the wick, using metal tweezers, to
   obtain a flame about 1 cm high.

6. Check that the lamp burns satisfactorily for about 15 seconds in position under the calorimeter. If it goes out, either increase the flow of air or adjust the height of the lamp relative to the calorimeter. Once these adjustments have been made, they should not be changed for the rest of the experiment. Extinguish the flame and put on the cap.

7. Weigh the spirit lamp, including the cap, as accurately as possible and record the mass in Results Table 13a.

8. Stir the water in the calorimeter and record its temperature, to the nearest 0.1 °C.

9. Put the lamp under the calorimeter and light it.

10. Slowly and continuously stir the water in the calorimeter and watch the thermometer. When the temperature has risen by about 10 °C, extinguish the flame and immediately replace the cap. Record the maximum temperature of the water.

11. Re-weigh the spirit lamp and cap and record the mass.

12. Without removing the calorimeter from the stand, and holding both together carefully, pour away the water.

13. If you have time, repeat the experiment to increase the accuracy of your calibration. This second run should be much quicker because you should not need to make any adjustments (i.e. start at step 7).

14. Before doing any calculations, repeat the experiment using as many of the other alcohols as you have time for and complete Results Table 13b. If you have to use the same spirit lamp, you will have to empty it, rinse it with the new alcohol, and fit a new wick (or dry the old wick). If time is limited, your teacher may suggest that you share your results with other students, or may give you some pre-determined results.

Results Table 13a	1st run	2nd run	
Molar mass of propan-1-ol, M			g mol^{-1}
Initial mass of spirit lamp + alcohol, m_1			g
Final mass of spirit lamp + alcohol, m_2			g
Mass of alcohol burned, $m_1 - m_2$			g
Amount of alcohol burned, $n = (m_1 - m_2)/M$			mol
Initial temperature of calorimeter			°C
Final temperature of calorimeter			°C
Temperature change, ΔT			K
$\Delta H_C^{\ominus}$ [propan-1-ol] (given)		-2017	kJ mol^{-1}
Heat released during the experiment, ΔH $= \Delta H_C^{\ominus}$ [propan-1-ol] × amount burned $= -2017$ kJ mol^{-1} × n			kJ
Heat required for a rise in temperature of 1 K $= \dfrac{\Delta H}{\Delta T}$ $= C$, the calorimeter calibration factor			kJ K^{-1}
Average value of C			kJ K^{-1}

Results Table 13b	C_4H_9OH	$C_5H_{11}OH$	$C_6H_{13}OH$	$C_7H_{15}OH$	$C_8H_{17}OH$
Molar mass, M/g mol^{-1}					
Initial mass of lamp/g					
Final mass of lamp/g					
Mass of alcohol burned/g					
Amount burned, n/mol					
Initial temperature/$^{\circ}$C					
Final temperature/$^{\circ}$C					
Temperature change, ΔT/K					
$\Delta H_C = \dfrac{C \times \Delta T}{n}$/kJ mol^{-1}					

Questions

1. Calculate the <u>difference</u> between the values of $\Delta H_C^{\ominus}$ for propan-1-ol and butan-1-ol, using both your experimental values and data book values.

2. Repeat the calculation for each adjacent pair of alcohols in the series and complete Table 13c.

   Table 13c

Pair of alcohols	Difference in $\Delta H_C^{\ominus}$/kJ mol^{-1}	
	Experiment	Data book
Propan-1-ol/butan-1-ol		
Butan-1-ol/pentan-1-ol		
Pentan-1-ol/hexan-1-ol		
Hexan-1-ol/heptan-1-ol		
Heptan-1-ol/octan-1-ol		

   Average difference = - - - kJ mol^{-1}

3. Relate the average difference to the bond energy terms for C—C and C—H.

4. Why is the difference in $\Delta H_C^{\ominus}$ fairly constant?

5. Why is the difference in $\Delta H_C^{\ominus}$ not precisely constant?

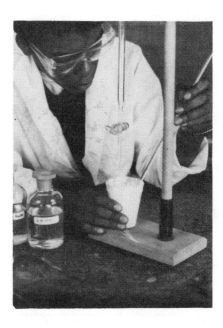

EXPERIMENT 14
A thermometric titration

Aim

The purpose of this experiment is to determine
the concentrations of two acids, hydrochloric
acid, HCl, and ethanoic acid, CH_3CO_2H, by thermometric titration; and having done that, to
calculate the enthalpy change for each reaction -
the enthalpy change of neutralization.

Introduction

You titrate both hydrochloric acid and ethanoic acid in turn with a standardized solution of sodium hydroxide and record the temperatures of the mixtures
during the course of the titrations. In each case a plot of temperature against
time will enable you to determine the maximum temperature rise, from which you
calculate both the concentration of the acid and the enthalpy change of
neutralization.

Requirements

safety spectacles
pipette, 50.0 cm³
pipette filler
expanded polystyrene cup
sodium hydroxide solution, 1 M NaOH (standardized) _ _ _ _ _ _ _ _ _ _ _ _ _ _
thermometer, 0-50 °C (in 0.1 °C)
burette, 50.0 cm³
filter funnel, small
hydrochloric acid, ~ 2.0 M HCl
ethanoic acid, ~ 2.0 M CH_3CO_2H

Hazard warning

Sodium hydroxide is very corrosive. Therefore you must:

USE THE PIPETTE FILLER SUPPLIED

WEAR SAFETY SPECTACLES

Procedure

Titration of hydrochloric acid with standard sodium hydroxide solution

1. Using a pipette and filler, transfer 50.0 cm³ of NaOH solution
   into the polystyrene cup. Allow to stand for a few minutes.

2. Record the temperature of the solution.

3. From a burette, add 5.0 cm³ of HCl solution to the cup.

4. Stir the mixture with the thermometer and record its temperature.

5. Add successive 5.0 cm³ portions of HCl solution stirring the mixture and
   recording its temperature after each addition.

6. Record your results in a copy of Results Table 14a. Stop after the
   addition of 50.0 cm³ of acid.

Titration of ethanoic acid with standard sodium hydroxide solution

7. Follow the same procedure as you did for the titration of HCl, except that you use ethanoic acid in the burette. When filling the burette, remember to use correct rinsing procedures. If in doubt, ask your teacher.

8. Record your results in a copy of Results Table 14b.

Results Table 14a Titration of hydrochloric acid

Volume added/cm³	0.0	5.0	10.0	15.0	20.0	25.0	30.0	35.0	40.0	45.0	50.0
Temper-ature/°C											

Results Table 14b Titration of ethanoic acid

Volume added/cm³	0.0	5.0	10.0	15.0	20.0	25.0	30.0	35.0	40.0	45.0	50.0
Temper-ature/°C											

Calculation

1. Plot temperature (y-axis) against volume of acid added (x-axis) for each acid on the same graph.

2. Extend the straighter portions of the curves near the top, as shown in Fig. 14. The point at which they meet corresponds to both the volume of acid required for neutralization and to the maximum temperature.

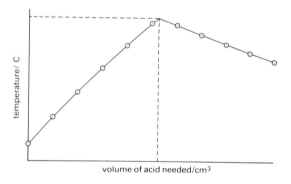

Fig. 14.

3. Calculate the concentration of each of the acids.

4. From the maximum temperature rise, determine the quantity of energy released in each titration. Assume that the specific heat capacity of the solutions is the same as that for water, 4.18 kJ kg⁻¹ K⁻¹ and that the heat capacity of the cup is zero.

5. Calculate the standard enthalpy change of neutralization for each reaction.

Questions

1. The enthalpy change of neutralization for a very dilute strong acid (i.e. an acid which is completely ionized in solution) reacting with a very dilute strong base is constant at -57.6 kJ mol⁻¹ where mol⁻¹ refers to one mole of water produced. Why is the value constant?

2. Experimental results for hydrochloric acid are usually a little less negative than -57.6 kJ mol⁻¹. Suggest two reasons for this.

3. Ethanoic acid is a weak acid, i.e. it is not completely ionized in solution. Suggest a reason why the heats of neutralization for reactions involving weak acids and/or weak bases are always less negative than for strong acids and bases.

EXPERIMENT 15
Making models of two metallic structures

Aim

The aim of the experiment is to make and
compare models of two structures commonly
found in metals - cubic close-packing and
hexagonal close-packing. In the course of
the construction, the concepts of close-
packing, co-ordination number, and unit cell
are illustrated and clarified.

Introduction

The model-building experiment is in the form of a revealing exercise. You
should cover with a sheet of paper that part of the page below the next ruled
line until you have followed the instruction and answered the question above
the line.

You will use expanded polystyrene spheres to represent atoms of metals and
stick them together with small blobs of Blu-tak or a similar demountable
adhesive. To get maximum rigidity, you should use adhesive on every contact
(except where you are told not to do so!) and press the spheres firmly
together so that they are as close as possible.

Do not dismantle any structure until the end except where you are told.

Requirements

31 expanded polystyrene spheres
Blu-tak or similar adhesive

Procedure

1. Put one sphere on a flat surface and surround it with as many others as
    you can fit in the same plane, i.e. all must touch both the flat surface
    and the central sphere. Stick the surrounding spheres together but not
    to the central sphere (you will need to remove it later).

Q1. How many spheres touch the central sphere, and what shape would be out-
    lined by joining their centres?

A1. Six spheres touch the central sphere, outlining a regular hexagon.

2. You have just constructed part of a close-packed layer or plane. Satisfy
    yourself that this arrangement can be continued to infinity in any direc-
    tion by adding more spheres (without sticking) around any of the outer
    ones.

Q2. What is the co-ordination number of any sphere in an infinite close-
    packed layer?

A2. Six. Note that this only refers to a 2-dimensional structure. You will
    work out the co-ordination number for a 3-dimensional structure later.

37

3. Stick three more spheres to your hexagon to make a triangular layer. Now you will extend close-packing to three dimensions. Try a single sphere, without adhesive, in different positions as part of a second layer to help you answer Q3 and Q4.

Q3. What is the maximum number of spheres in the first layer that can be touched by a single sphere in the second layer?

A3. Three.

Q4. How many sites are there where you can add a single sphere in the second layer so that it touches three others in the first layer?

A4. Nine. These are shown by crosses in Fig. 15.

Fig. 15.

5. Add more spheres to make a close-packed second layer - do not stick them together yet.

Q5. How many of the nine sites can you actually use to make a close-packed layer?

A5. Either six or three. In an infinite layer, half the sites can be used, corresponding to alternate crosses in Fig. 15.

6. Consider both ways of adding the second layer - adding three spheres or six spheres - and answer the questions for each of the ways.

Q6. (a) Is the second layer close-packed?

     (b) Is it close-packed with respect to the first layer?

     (c) If the layers were extended in all directions, could you distinguish between the two ways of adding the second layer?

A6. (a) Yes, for both ways. (b) Yes, for both ways.

     (c) No, the two ways of adding the second layer are indistinguishable.
          It is only the fact that you are looking at a small part of the
          structure that makes them seem different.

7. Arrange your second layer to have six close-packed spheres and stick them together as a triangle. This is best done on a flat surface - then lift the complete layer back into position. Do not stick the layers together.

Now consider the different ways of adding a third layer.

Q7. How many close-packing sites are there? Can you use them all?

A7.   There are four sites. You can use three or one. In an extended layer, <u>half</u> the sites can be used.

8.   Look carefully at the four sites from a position vertically above each one, so that you look through to the first layer.

Q8.   Is there any difference between the sites? If so, what difference can you see?

A8.   The central site is directly above a sphere in the first layer. The other three sites are directly above holes in the first layer.

9.   Use the three outer sites to make a third layer. Stick the spheres together but not to the second layer.

Now stick another sphere in position to make a fourth layer. You have now constructed a model of part of the structure adopted by many metals, e.g. copper, silver and gold. This structure is known by two names - *abc* close-packing or cubic close-packing. It is sometimes called a face-centred cubic structure, which is correct but not precise because there are other face-centred cubic structures.

Q9.   Why do you think the structure is known as *abc* close-packing? (You have answered this in a different way in Q8.)

A9.   The third layer is not directly above the first, but the fourth layer is. The first three layers are referred to as *a*, *b* and *c*, while the fourth is regarded as another *a* layer. The sequence continues *abcabcabcabc*....

10.   It is not so easy to see why the structure is called cubic, although you may perhaps know that the shape of your pyramid is a regular tetrahedron which fits into a cube with its points at four of the eight corners, as shown in Fig. 16.

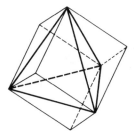

Fig. 16.

To show the cubic nature of the structure more clearly, take the second layer of your tetrahedron, six spheres still firmly stuck together, and place it on a flat surface. Stick one more sphere in the central site so that you have a two-layer structure of 7 spheres.

Make another identical 7-sphere structure by adding three spheres to the third and fourth layers removed from the tetrahedron.

Hold one structure in each hand by means of the single sphere and bring the two six-sphere triangles together. Then rotate one triangle till it fits into the other, using three close-packing sites in each triangle.

Q10.   Describe the structure you now have in a few words - what is its overall shape and how do the spheres fit into it?

A10.   It is a cube, each face of which has a sphere at each corner and one at the centre touching the other four - hence the name 'face-centred cubic.'

Q11. Do the faces of your cube represent close-packed layers?

A11. No. The co-ordination number is 4 and not 6.

Q12. Where are the close-packed layers in the cube?

A12. Along planes joining any diagonal of a face with <u>one</u> other corner. The shaded spheres in Fig. 17 make one such layer.

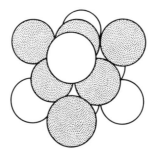

Fig. 17.

Q13. What is the co-ordination number of any sphere in the extended structure?

A13. Twelve. Six touch it in one close-packed layer, three in the layer above, and three in the layer below.

14. Turn the cube and stand it on one corner so that the close-packed planes are horizontal. There are four such planes, containing 1, 6, 6 and 1 sphere. Check that the sequence is *abca* as in A9.

Another way of seeing that the face-centred cube you have made has the same structure as the pyramid you built at first is to remove the central sphere from the original first layer and stand the cube with a corner sphere in the hole in such a way that the original pyramid shape can be seen, but with an extra sphere projecting from three of its faces.

Draw an outline cube and draw small blobs (ideally they should be just points) in positions corresponding to the centres of the spheres in your model structure. Compare your drawing with drawings in your textbook. The drawing represents the unit cell of the cubic close-packed structure since the whole structure can be built up by repeating it.

Q14. However, the face-centred cube you have made from spheres is not, strictly speaking, a unit cell. Can you see why not?

A14. The stacking of identical cubes side-by-side would <u>not</u> repeat the structure, unless the spheres are regarded as being <u>shared</u> between neighbouring cubes. In Fig. 18, there is half a sphere at the centre of each face, the other half belonging to a neighbouring cube. Similarly, spheres with their centres at the corners of the cube are shared between eight neighbouring cubes.

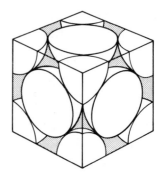

Fig. 18.

15. Now you will investigate the structure made by using the second way of arranging the third layer.

Return half of your face-centred cube to the hole in the first layer so that you have two triangular layers just as you did at the beginning. You recall that to make the cubic close-packed structure, the third layer spheres used the outer three close-packing sites. This time, place a sphere in the central site and stick six more spheres to it to make a third layer.

Q15. How do the positions of the third layer spheres relate to those in the first layer?

A15. Each sphere is vertically above a sphere in the first layer.

16. You have now made a model of part of the structure adopted by a number of other metals, e.g. magnesium and calcium. This structure is also known by two names - *aba* close-packing or hexagonal close-packing.

Q16. Why is it called *aba* close-packing?

A16. The third layer is directly above the first. Compare this with cubic close-packing where the fourth layer is directly above the first.

17. The hexagonal nature of the structure is best seen by removing the three corner spheres from the first layer and making these three the second layer (first filling the hole in the first layer). Finally, replace the third layer directly above the first, so that the overall hexagonal shape is apparent.

Compare your model with drawings in your textbooks and with a permanent model, if one is available. You should be able to draw an outline of the hexagonal close-packed unit cell using only dots and lines, as you did for cubic close-packing. Remember that your model is not strictly a unit cell unless spheres are shared with neighbouring cells.

18. Before you dismantle your models, look at them again, side by side, and make sure that you can:

(a) pick out the close-packed layers;

(b) see that the overall co-ordination number is 12 in both cases;

(c) draw unit cells, representing the centres of spheres by dots.

EXPERIMENT 16
Recognising ionic, covalent and metallic structures

Aim

The aim of this experiment is to examine some unknown substances and to determine their structures (as far as possible) using the minimum number of tests.

Introduction

A number of tests are suggested, as shown in Results Table 16. You should devise your own simple procedures, using only the apparatus provided.

Do not necessarily perform every test on each substance, but aim at getting the maximum information from the minimum number of tests. The result of one test may suggest which test to do next. In some cases, it may not be possible to come to a definite conclusion. In estimating melting points to the nearest 100 °C, you should note that the maximum temperature of an ordinary Bunsen burner flame is about 800 °C.

Requirements

safety spectacles	beaker, 100 cm³
6 test-tubes in rack	battery and lamp in holder
6 ignition tubes	2 carbon electrodes
test-tube holder	3 connecting leads with crocodile clips
Bunsen burner and protective mat	unknown substances, in bottles labelled A-G

Results Table 16

	A	B	C	D	E	F	G
Appearance							
Estimate of boiling-point							
Solubility in water							
Conductivity of solution							
Conductivity of solid							
Action of dilute HCl							
Structure							

Questions

1. What further tests would probably help to identify the structures which you could not identify in the experiment?

2. Why is it more difficult to recognise a powdered metal than a solid lump?

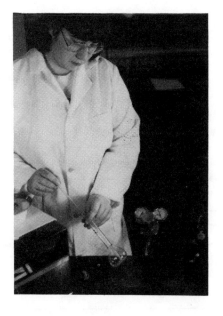

EXPERIMENT 17
Determining the molar mass of a gas

Aim

The purpose of this experiment is to measure the volume and mass of a sample of carbon dioxide, and to use these values to determine the molar mass.

Introduction

You weigh a clean dry flask full of air and then full of carbon dioxide. By filling the flask with water and reweighing, you can find its volume. Knowing the density of air, you calculate the mass of air filling the flask and use it to find the mass of the empty flask, and hence the mass of carbon dioxide. You then use the ideal gas equation to determine the molar mass of carbon dioxide.

Requirements

volumetric flask, 100 cm^3, dry, with stopper
balance(s) capable of taking volumetric flask and of weighing -
  (a) up to 100 g with accuracy of 0.001 g
  (b) up to 200 g with accuracy of 0.1 g
carbon dioxide cylinder or generator
delivery tube, glass, 20-30 cm long
rubber tubing, 30-90 cm, to connect gas cylinder to delivery tube
thermometer, 0-100 ^{o}C
access to barometer (or telephone number of local meteorological office)

Procedure

1. Get instructions from your teacher on how to operate the gas cylinder - there will be some valves you must not touch. Alternatively, you can use a simple carbon dioxide generator, provided you purify the gas from acid spray and dry it before use. Again, ask your teacher.

2. Weigh the dry volumetric flask together with its stopper to the nearest 0.001 g. Enter the mass in Results Table 17a.

3. Remove the stopper, insert the glass delivery tube from the carbon dioxide cylinder or generator so that it reaches the bottom of the flask, and open the valve so that gas passes through for at least one minute. Keep the flask upright throughout.

4. Slowly remove the delivery tube, quickly close the flask with the stopper and close the valve on the cylinder or generator.

5. Weigh the flask with stopper again to the nearest 0.001 g.

6. Repeat steps 3, 4 and 5, and check that there is no further change in mass (i.e. that the carbon dioxide has indeed displaced all the air from the flask). If this is not the case, repeat these steps again - and then yet again, if necessary, until the mass is constant.

43

7. Fill the flask with water and insert the stopper, so that excess water is pushed out. Dry the outside of the flask and weigh it, full of water, on a robust balance to the nearest 0.1 g.

8. Note room temperature and atmospheric pressure.

9. Complete the results table below.

Results Table 17a

Mass of flask filled with air	g
Mass of flask filled with CO_2	g
Mass of flask filled with water	g
Room temperature	^{o}C
Atmospheric pressure	mmHg
Density of air under conditions of experiment	g cm^{-3}

The following table gives values for the density of the air under various conditions of temperature and pressure. If your conditions do not correspond to any of those quoted, you should estimate the appropriate value.

Table 17b   Density of air (g cm^{-3}) at different temperatures and pressures.

	15 ^{o}C	17 ^{o}C	19 ^{o}C	21 ^{o}C	23 ^{o}C	25 ^{o}C
740 mmHg	0.00119	0.00119	0.00118	0.00117	0.00116	0.00115
750 mmHg	0.00121	0.00120	0.00119	0.00119	0.00118	0.00117
760 mmHg	0.00123	0.00122	0.00121	0.00120	0.00119	0.00119
770 mmHg	0.00124	0.00123	0.00123	0.00122	0.00121	0.00120
780 mmHg	0.00126	0.00125	0.00124	0.00123	0.00122	0.00122

Calculation

You need to calculate the mass of carbon dioxide from your experimental results, before using the ideal gas equation in the form $pV = mRT/M$. The steps in the calculation are as follows.

Calculate   1. the volume of the flask (from the mass and density of water);

2. the mass of air in the flask;

3. the mass of the empty stoppered flask (i.e. with no air in it);

4. the mass of carbon dioxide in the flask;

5. the molar mass of carbon dioxide.

Questions

1. What value does the experiment give for the <u>relative</u> molecular mass of CO_2?

2. Calculate the density of CO_2 at s.t.p. from your results.

3. In step (4) why were you told to remove the delivery tube slowly?

4. Why is a less accurate balance adequate for weighing the flask full of water?

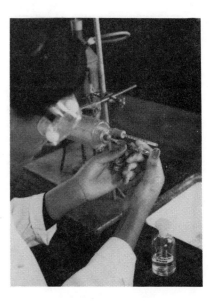

EXPERIMENT 18

Determining the molar mass of a volatile liquid

Aim

The aim of the experiment is to determine the
molar mass of 1,1,1-trichloroethane, CH_3CCl_3,
at the temperature of boiling water and at
atmospheric pressure. The same method can be
used for other liquids which boil at a temperature below 80 °C, but we have chosen this
one because it is non-toxic and non-flammable.

Introduction

You obtain the mass of a sample of the liquid by weighing a small hypodermic
syringe before and after injection into a large gas syringe. The large
syringe is heated in a steam jacket (see Fig. 19) and you measure the volume
of the vapour at the temperature of condensing steam. Finally, you apply the
ideal gas equation as before to calculate the molar mass.

Requirements

safety spectacles
100 cm³ gas syringe, glass
self-sealing rubber cap for gas syringe
steam jacket for gas syringe
thermometer, 0 °C to 100 °C, to fit steam jacket
Bunsen burner, tripod, gauze and bench mat
steam generator, with safety tube
2 cm³ hypodermic syringe, glass, and needle
1,1,1-trichloroethane, CH_3CCl_3
filter paper
self-sealing silicone rubber to make temporary seal for needle
balance (accuracy 0.001 g)
access to barometer (or telephone number of local meteorological office)

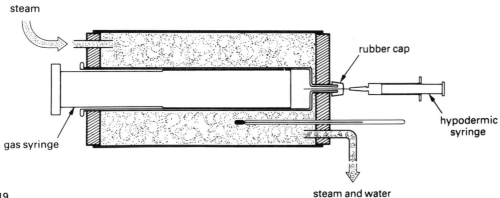

Fig. 19.

Procedure

1. Place the gas syringe into the steam jacket and draw in about 5 cm³ air
    before sealing it with the rubber cap.

2. Pass steam through the steam jacket until the temperature reading and
    the volume of air in the syringe reach steady values. You can begin
    the next step while you are waiting for this steady state to be
    established.

3. Draw about 1 cm³ of 1,1,1-trichloroethane into the hypodermic syringe
    through the needle, rinse the syringe with the liquid and expel it into
    the sink. Draw in another 1 cm³ of liquid and, holding the syringe
    vertically with the needle uppermost, slowly push in the piston till
    every bubble of air is expelled and a few drops of liquid emerge.

4. Dry the outside of the needle with filter paper and seal it with a
    small piece of silicone rubber.

5. Weigh the hypodermic syringe with its cap and liquid contents and record
    the mass in Results Table 18. Keep the syringe horizontal and avoid
    touching the piston or warming the barrel with your hand, either of
    which could result in loss of liquid.

6. When the temperature and volume of the air in the gas syringe are
    constant, record the volume of air and, with steam still passing through
    the jacket, push the hypodermic needle through its own seal and through
    the rubber seal of the gas syringe so that its tip projects well into
    the air space. Inject about 0.2 cm³ of the liquid into the gas syringe
    (see Fig. 20).

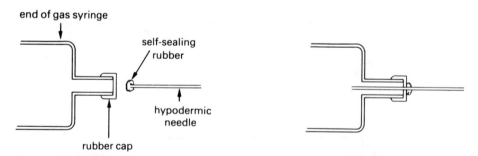

Fig. 20.

7. Withdraw the needle into its own self-sealing cap, and re-weigh the
    syringe, cap and contents immediately so that no more liquid escapes.
    Record the mass.

8. Make sure, by twirling it, that the piston in the gas syringe can move
    freely so that the pressure inside is the same as atmospheric pressure,
    which should also be recorded.

9. Watch the temperature, and the volume of air and vapour in the syringe,
    until both reach steady values. Record these steady values.

10. Remove the cap from the gas syringe and push the piston in and out several times to expel the vapour.

11. If you, or another student, wish to use the apparatus again immediately, leave the steam generator going. Otherwise turn off the Bunsen burner.

Results Table 18

Mass of hypodermic syringe and liquid before injection	g
Mass of hypodermic syringe and liquid after injection	g
Temperature of vapour	°C
Atmospheric pressure	mmHg
Volume of air in syringe	cm³
Volume of air and vapour in syringe	cm³

Calculation

From your results, calculate the molar mass of 1,1,1-trichloroethane by using the ideal gas equation in the same way as in Experiment 17.

Questions

1. What value does the experiment give for the relative molecular mass of 1,1,1-trichloroethane?

2. What might happen if the hypodermic needle were shorter than the nozzle of the gas syringe, and what effect would this have on your final results?

3. Calculate the molar mass from your experimental results in a different way, using the known value for molar volume at s.t.p.

4. The results obtained using the alternative method is very slightly different. Can you suggest why this is so?

EXPERIMENT 19

Determining the molar mass of
a gas by effusion

Aim

The purpose of this experiment is to apply
Graham's law to determine the molar mass
of domestic gas.

Introduction

You fill a gas syringe with hydrogen, and allow it to escape through a
small pin hole under the weight of the piston. You measure the rate of
escape and repeat the experiment using domestic gas. Assuming the molar mass
of hydrogen, you use Graham's law to calculate the molar mass of domestic gas.

Requirements

small piece of aluminium foil
quick-setting glue
three-way tap and connector to fit syringe
gas syringe, 100 cm³
retort stand and clamp
needle or pin
stopclock or stopwatch (preferably to 0.1 sec)
hydrogen cylinder or generator with rubber delivery tube
tubing to gas tap

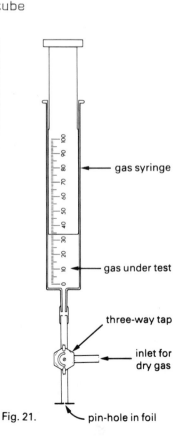

Hazard warning

Hydrogen and domestic gas are flammable and
they can form explosive mixtures with air.

THERE MUST BE NO FLAMES IN THE LABORATORY
DURING THIS EXPERIMENT.

Procedure (steps 1 to 3 may already be done for you)

1. Cut a piece of smooth aluminium foil about 1 cm
   square and lay it on a flat surface, ready to
   be glued to the three-way tap. Do not pierce
   the foil yet.

2. Apply a little glue to the flat end of one of
   the tubes of the three-way tap. There must be
   enough to form a complete seal between the foil
   and the glass, but not so much as to close the
   aperture.

3. Lower the three-way tap vertically on to the
   foil, press gently together, and leave until the
   glue has dried.

gas syringe

gas under test

three-way tap

inlet for
dry gas

pin-hole in foil

Fig. 21.

4. Check that the syringe piston moves freely (it must not be greased), attach the three-way tap and clamp the syringe carefully as shown in Fig. 21. Turn the tap so that there is a channel between side tube and syringe (position A - Fig. 22) and fill the syringe with air.

5. Turn the tap so that the side tube is closed but gas can pass between syringe and foil (position B - Fig. 23).

6. Watch the volume reading of the syringe for half a minute - it should not change. If the volume does change, air must be escaping from the three-way tap, or from the connector, or from the joint between foil and glass. Check these in turn, regreasing the tap if necessary, until you are satisfied that the apparatus does not leak.

   (There is always some leakage between the piston and the syringe, but in a good syringe this will be so slow as to be negligible over half a minute. If it is not negligible, you can dispense with the three-way tap and the foil, seal the nozzle with a bung, and use the gap between piston and syringe for the effusion!)

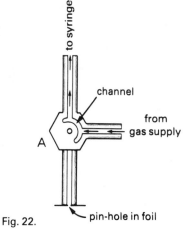

channel

from gas supply

A

pin-hole in foil

Fig. 22.

7. With the tap still in position B, carefully pierce the foil with a needle, making a very small hole at first. Watch the rate of fall of the piston and increase the size of the hole, if necessary, until the piston falls at a rate equivalent to about 1 cm³ air expelled each second. This sets the size of hole for the complete experiment - do not touch it again!

8. Turn the tap to position A and expel the air from the syringe through the side tube.

9. Check with your teacher on any instructions for using the hydrogen cylinder or generator (NO FLAMES NEAR!) and obtain a slow flow of gas through the delivery tube before connecting it to the side tube of the three-way tap.

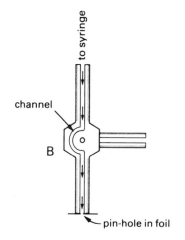

channel

B

pin-hole in foil

Fig. 23.

10. Fill the syringe to about 50 cm³, remove the delivery tube and expel the gas from the syringe to ensure that no air remains. Refill to about 75 cm³, turn off the cylinder or generator and turn the tap to position B.

11. Allow the piston to fall to the 60 cm³ mark and start the stopclock at the moment it passes the mark.

12. Stop the clock when the piston passes the 10 cm³ mark and record the time in Results Table 19.

13. Repeat steps 9 to 12 at least twice. The times should not vary by more than 10% - if they do, check again for leaks, particularly from the tap, and for a sticking piston. Repeat, if necessary, to obtain reproducible results.

14. Expel the hydrogen from the syringe, obtain a slow flow of domestic gas through a rubber tube from a gas tap and fill the syringe, as before, repeating steps 10 to 13, as necessary, to complete the results table.

Results Table 19

	1	2	3	4	Mean
Time for effusion of 50 cm³ of hydrogen/s					
Time for effusion of 50 cm³ of domestic gas/s					

Calculation

Use your results to calculate the molar mass of domestic gas using this formulation of Graham's law:

$$\frac{M(H_2)}{M(gas)} = \left(\frac{t(H_2)}{t(gas)}\right)^2$$

Questions

1. What is the main constituent of domestic gas? Is your result for the molar mass consistent with your answer?

2. Does your result suggest that impurities in domestic gas have molar masses greater or less than the molar mass of the main constituent? What might these impurities be?

3. Why were you told to allow the piston to fall from the 75 cm³ mark to the 60 cm³ mark before starting the stop-clock?

4. What difference would you expect if you repeated the experiment with the same apparatus and the same pinhole at a higher temperature? Explain.

5. What difference would you expect if you repeated the experiment with the same apparatus at the same temperature but with a larger hole?

EXPERIMENT 20

The effect of concentration changes
on equilibria

Aim

The purpose of this experiment is to find
out how a system in equilibrium responds
to a change in concentration of components
in the mixture.

Introduction

Iron(III) ions and thiocyanate ions react in solution to produce
thiocyanatoiron(III), a complex ion, according to the equation:

$$Fe^{3+}(aq) + SCN^-(aq) \rightleftharpoons Fe(SCN)^{2+}(aq)$$

| pale yellow | colourless | blood-red |

The colour produced by the complex ion can indicate the postion of
equilibrium.

Requirements

safety spectacles
4 test-tubes and test-tube rack
2 teat-pipettes
distilled water
potassium thiocyanate solution, 0.5 M KSCN
iron(III) chloride solution, 0.5 M FeCl$_3$
ammonium chloride, NH$_4$Cl
spatula
glass stirring rod

Procedure

1. Mix together one drop of 0.5 M iron(III) chloride solution and one
   drop of 0.5 M potassium thiocyanate solution in a test-tube and add
   about 5 cm^3 of distilled water to form a pale orange-brown solution.

2. Divide this solution into four equal parts in four test-tubes.

3. Add one drop of 0.5 M iron(III) chloride to one test-tube.
   Add one drop of 0.5 M potassium thiocyanate to a second.

4. Compare the colours of these solutions with the untouched samples.
   Enter your observations in a copy of Results Table 20.

5. Add a spatula-full of solid ammonium chloride to a third test-tube
   and stir well. Compare the colour of this solution with the
   remaining tube and note your observation.

   Ammonium chloride removes iron(III) ions from the equilibrium by forming
   complex ions such as FeCl$_4^-$. A possible reaction is:

   $$Fe^{3+}(aq) + 4Cl^-(aq) \rightleftharpoons FeCl_4^-(aq)$$

   The effect is to reduce the concentration of iron(III) ions.

Interpretation of results

Having made three observations, suggest a cause for each colour change (in terms of the concentrations of the coloured species) and then suggest what can be inferred about a shift in the position of equilibrium.

If a pattern has emerged, then you can make a prediction based on the results of the experiment.

Results Table 20

Change	Observation	Cause	Inference
$[Fe^{3+}]$ increased			
$[SCN^-]$ increased			
$[Fe^{3+}]$ decreased			

Questions

1. How would the position of equilibrium be affected by increasing the concentration of $FeSCN^{2+}$?

2. For each imposed change show how the shift in equilibrium position conforms to Le Chatelier's principle.

EXPERIMENT 21
Determining an equilibrium constant

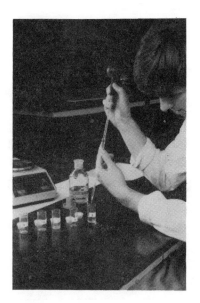

Aim

The purpose of this experiment is to
calculate the equilibrium constant for
the reaction:

$CH_3CO_2C_2H_5(l) + H_2O(l)$
ethyl ethanoate   water

$\rightleftharpoons C_2H_5OH(l) + CH_3CO_2H(l)$
    ethanol   ethanoic acid

Introduction

The reaction between ethyl ethanoate and water is very slow. However, by
using a catalyst, dilute hydrochloric acid, equilibrium can be attained in
about forty-eight hours.

In part A of the experiment you prepare, in sealed containers, mixtures
containing different proportions of the two reactants. To each mixture you
add a fixed amount of dilute hydrochloric acid as a catalyst.

In part B, after the mixtures have reached equilibrium at room temperature,
you analyse each one by titration with sodium hydroxide. Part of the added
sodium hydroxide reacts with the catalyst; the rest indicates the amount of
ethanoic acid in the equilibrium mixture.

Finally, from the starting amounts and the amount of ethanoic acid produced,
you calculate the equilibrium concentrations of all four components and use
them to determine the equilibrium constant.

Requirements - Part A

safety spectacles
5 specimen tubes with well-fitting caps
labels for tubes and stoppers
access to a balance (sensitivity ± 0.01 g or better)
pipette, 5 cm³, and safety filler
dilute hydrochloric acid, 2 M HCl
2 measuring cylinders, 10 cm³ (one must be dry)
ethyl ethanoate, $CH_3CO_2C_2H_5$ —
distilled water

Procedure - Part A

1. Label five specimen tubes with your name and the date. Number them 1A,
   . 1B, 2, 3 and 4. Number the stoppers too, so that they do not get
   misplaced.

2. Weigh each tube, with its stopper, and record the masses in a copy of
   Results Table 21a.

3. Using a pipette and safety filler, carefully add 5.0 cm³ of 2 M
   hydrochloric acid to each tube, replacing the stoppers as you go. The
   volume of acid must be precisely the same in each tube; measure it as
   carefully as you can. If you think you have made a mistake, wash out the
   tube, dry it and start again.

4. Weigh each stoppered tube in turn and record the masses.

5. Select a dry measuring cylinder, and use it to add to tubes 2, 3 and 4 the volumes (approximate) of ethyl ethanoate shown in Results Table 21a, again replacing the stoppers as you go.

6. Weigh the stoppered tubes 2, 3 and 4. Record the masses.

7. From a second measuring cylinder, add to tubes 3 and 4 the volumes (approximate) of distilled water shown in Results Table 21a, again replacing the stoppers as you go.

8. Weigh the stoppered tubes 3 and 4. Record the masses.

9. Gently shake the tubes and set them aside for at least 48 hours. During this time, arrange to shake the tubes occasionally.

Results Table 21a

Tube number	1A	1B	2	3	4
Mass of empty tube/g					
Volume of HCl(aq) added/cm³	5.0	5.0	5.0	5.0	5.0
Mass of tube after addition/g					
Volume of ethyl ethanoate added/cm³	-	-	5.0	4.0	2.0
Mass of tube after addition/g					
Volume of water added/cm³	-	-	-	1.0	3.0
Mass of tube after addition/g					
Mass of ethyl ethanoate added/g					
Mass of HCl(aq) added/g					
Mass of water added/g					

In addition to the two tubes containing only hydrochloric acid, you now have three tubes containing different amounts of ethyl ethanoate and water (tube 2 has water from the acid). When these mixtures have reached equilibrium, you can analyse them in part B of the experiment.

Before you begin part B, revise the technique of titration (Experiment 3).

Requirements - Part B

safety spectacles
5 conical flasks, 250 cm³
wash-bottle of distilled water
phenolphthalein indicator
burette, stand and white tile
small funnel
sodium hydroxide solution, 1 M NaOH (standardized) — — — — — — — — — —

Procedure - Part B

1. Rinse and fill a burette with standardized sodium hydroxide solution.

2. Carefully pour the contents of tube 1A into a conical flask. Rinse the tube into the flask three times with distilled water.

3. Add 2-3 drops of phenolphthalein indicator solution and titrate the acid against sodium hydroxide solution. Record your burette readings in a copy of Results Table 21b.

4. Repeat steps 2 and 3 for each of the other tubes in turn. Remember that tube 1B should require the same volume of alkali as tube 1A, but the others should require a little more. Complete Results Table 21b.

Results Table 21b

Solution in flask	Equilibrium mixture				
Solution in burette	Sodium hydroxide mol dm^{-3}				
Indicator	Phenolphthalein				
Tube number	1A	1B	2	3	4
Final burette reading					
Initial burette reading					
Titre/cm^3					

Calculations

The calculations seem complex, but consist of several very simple steps. The flow scheme below summarises the procedure; refer to it as you work through so that you can see the purpose of each step.

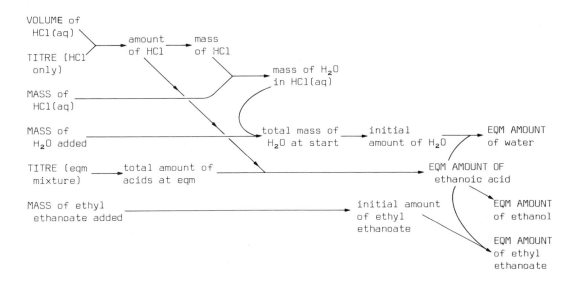

1. From the average titre for tubes 1A and 1B (or the better titre if you think one was inaccurate), calculate the amount of hydrochloric acid catalyst you added to each of the five tubes. Remember that hydro-chloric acid and sodium hydroxide react in equimolar amounts and that

        amount = concentration x volume.

    Record this amount in each column of a copy of Results Table 21c.

2. From the titres for tubes 2, 3 and 4 in turn, calculate and record the total amount of acid (hydrochloric and ethanoic) in each mixture.

3. Calculate, by subtraction, the amount of ethanoic acid in each equilibrium mixture. This is the first of the four quantities you need to substitute in the equilibrium law expression. Complete the third row of Results Table 21c.

4. The equation for the equilibrium system shows that the amount of ethanol produced is equal to the amount of ethanoic acid produced. You now have the second of the four quantities you need. Complete the fourth row of your table.

5. From the data in Results Table 21a, calculate the amount of ethyl ethanoate added to each tube.

$$\text{amount of ethyl ethanoate} = \frac{\text{mass after addition - mass before addition}}{\text{molar mass}}$$

Complete the fifth row of your table.

6. Calculate and record the equilibrium amount of ethyl ethanoate in each tube, using the relationship:

$$\text{eqm amount of } CH_3CO_2C_2H_5 = \text{initial amount of } CH_3CO_2C_2H_5 - \text{eqm amount of } CH_3CO_2H$$

You can see from the chemical equation that the amount of ethyl ethanoate which reacts is equal to the amount of ethanoic acid produced.

You now have the third of the four quantities you need.

7. Calculate and record the mass of pure HCl in each mixture.

$$\text{mass} = \text{amount} \times \text{molar mass}$$

8. Calculate and record the mass of water in the aqueous HCl added to each tube. (You need to refer back to Results Table 21a.)

$$\text{mass of water} = \text{mass of HCl(aq)} - \text{mass of HCl}$$

9. Calculate and record the total amount of water initially in each mixture.

$$\text{initial amount of } H_2O = \frac{\text{mass in HCl(aq) + mass added}}{\text{molar mass}}$$

10. Calculate and record the equilibrium amount of water in each mixture

$$\text{eqm amount of } H_2O = \text{initial amount of } H_2O - \text{eqm amount of } CH_3CO_2H$$

11. Write an equilibrium law expression for the reaction and calculate three values of the equilibrium constant, K_c.

Results Table 21c

	Tube number	2	3	4
1.	Amount of HCl/mol			
2.	Total amount of acid at eqm/mol			
3.	Eqm amount of ethanoic acid/mol			
4.	Eqm amount of ethanol/mol			
5.	Initial amount of ethyl ethanoate/mol			
6.	Eqm amount of ethyl ethanoate/mol			
7.	Mass of pure HCl/g			
8.	Mass of water in HCl(aq)/g			
9.	Initial amount of water/mol			
10.	Eqm amount of water/mol			
11.	Eqm constant, K_c			

EXPERIMENT 22
Determining a solubility product

Aim

The purpose of this experiment is to
determine the solubility and solubility
product of calcium hydroxide.

Introduction

The equilibrium between solid calcium hydroxide and its ions in an aqueous
solution is

$$Ca(OH)_2(s) \rightleftharpoons Ca^{2+}(aq) + 2OH^-(aq)$$

The concentration of hydroxide ions can be determined by titration with
hydrochloric acid; the concentration of calcium ions can be calculated
from the titration result.

Requirements

safety spectacles
4 stoppered bottles, 250 cm³
labels for bottles
spatula
calcium hydroxide, solid, $Ca(OH)_2$
measuring cylinder, 100 cm³
distilled water
4 filter funnels, <u>dry</u>, with filter papers
4 conical flasks, $\overline{250}$ cm³
thermometer 0-100 ºC (± 1 ºC)
pipette, 25 cm³, and safety filler
burette and stand, white tile
small funnel
hydrochloric acid solution, 0.1 M - standardized
phenolphthalein indicator solution

Procedure

1. Into each of four bottles put about 2 g of powdered calcium hydroxide
    and about 100 cm³ of distilled water. Stopper securely.

2. Shake well for about a minute. Label each bottle with your name,
    experiment and date, and set aside for a day or more.

3. Rinse and fill the burette with standardized hydrochloric acid.

4. Filter the contents of one bottle, allowing the first 5 cm³ to run to
    waste and collecting the rest in a dry conical flask. (The first few
    cm³ are rejected because they are less concentrated in solute than the
    rest. The filter paper adsorbs solute until it attains equilibrium with
    the solution. Yet another equilibrium!)

Steps 5 and 6 should be done as quickly as possible (with due care!) and
with only the minimum shaking that will ensure mixing.

5. Rinse the pipette with the calcium hydroxide solution and transfer 25.0 cm^3 to a conical flask (this need not be dry).

6. Add two drops of phenolphthalein to the flask and titrate the solution until the pink colour just disappears. Record your burette readings in a copy of Results Table 22.

7. Repeat steps 4, 5 and 6 for the other three solutions.

8. Record the temperature.

Results Table 22

Solution in flask					mol dm^{-3}	cm^3
Solution in burette					mol dm^{-3}	
Indicator						
		Trial	1	2	3	4
Burette readings	Final					
	Initial					
Volume used/cm^3						
Mean titre/cm^3						

Calculation

1. Calculate the concentration of hydroxide ion in a saturated solution of calcium hydroxide.

2. From the equilibrium concentration of hydroxide ion calculate the equilibrium concentration of calcium ion.

3. Calculate the solubility of calcium hydroxide at the temperature of your experiment. Compare your result with the value listed in your data book.

4. Calculate the solubility product from:

   (a) your result,

   (b) the solubility of calcium hydroxide given in your data book.

5. Suggest a reason for the speed of working advised for steps 5 and 6 above. (Hint: slow working, with much shaking of the flask, gives a smaller titre.)

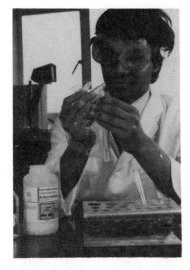

EXPERIMENT 23
Illustrating the common ion effect

Aim

The purpose of this experiment is to demonstrate an example of the application of Le Chatelier's principle.

Introduction

Although sodium chloride is quite soluble, we can use it to demonstrate the common ion effect. You should be able to relate your specific observations in this experiment to the general formulation of Le Chatelier's principle.

Requirements

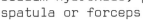

safety spectacles
2 test-tubes with corks, in a rack
sodium chloride solution, saturated, NaCl
hydrochloric acid, concentrated, HCl — — — — — — — — — — — — — — — —
teat-pipette
sodium hydroxide, pellets, NaOH — — — — — — — — — — — — — — — — — — —
spatula or forceps

Procedure

1. Carefully pour about 10 cm³ of saturated sodium chloride solution into each of two test-tubes. Do not transfer any solid.

2. To the first test-tube carefully add 4-5 drops of concentrated hydrochloric acid. Cork the tube, shake gently and set aside.

3. To the second test-tube add one pellet of sodium hydroxide.
   USE FORCEPS OR SPATULA TO HANDLE THE SODIUM HYDROXIDE.
   Cork the tube, shake gently and set aside.

4. Note any observations.

Questions

1. What did you see happen in each test-tube?

2. Interpret your observations in terms of Le Chatelier's principle.

3. In the saturated solution of sodium chloride, what is the concentration of each ion? (Use your data book.)

4. The concentration of concentrated hydrochloric acid is approximately 12 mol dm⁻³. Explain what happened to the sodium ions in solution when the hydrochloric acid was added.

5. Explain what happened to the chloride ions in solution when the sodium hydroxide pellet was added.

6. Why is no solubility product value given for sodium chloride in any data book?

EXPERIMENT 24
Distribution equilibrium

Aim

The purpose of this experiment is to determine the value of the distribution coefficient for the equilibrium that exists when ammonia is distributed between water and 1,1,1-trichloroethane.

Introduction

In this experiment you shake some ammonia solution with 1,1,1-trichloroethane to establish equilibrium, and then determine the concentration of ammonia in each solvent by titration. This enables you to calculate the distribution coefficient, K_D.

$$NH_3(tce) \rightleftharpoons NH_3(aq) \qquad (tce = 1,1,1\text{-trichloroethane})$$

$$K_D = \frac{[NH_3(aq)]}{[NH_3(tce)]}$$

Requirements

safety spectacles
measuring cylinder, 50 cm³
ammonia solution, 1 M NH₃
1,1,1-trichloroethane, CH₃CCl₃
separating funnel, 150 cm³
2 beakers, 100 cm³
pipette, 10 cm³, and safety filler
2 conical flasks, 150 cm³
wash-bottle of distilled water
methyl orange indicator solution
white tile
burette, 50 cm³, and stand
hydrochloric acid, 0.010 M HCl (standardized)
hydrochloric acid, 0.50 M HCl (standardized)

Procedure

1. Pour about 50 cm³ of ammonia solution into a separating funnel.

2. Pour about 50 cm³ of 1,1,1-trichloroethane into the same separating funnel.

3. Holding the tap firmly in position with one hand and the stopper with the other, shake the separating funnel vigorously for about ten seconds. Release the pressure inside by loosening the stopper for a moment.

4. Continue shaking for about half a minute, releasing the pressure every ten seconds. Set aside until two layers separate.

5. Transfer the lower organic layer to a beaker. Using a <u>dry</u> pipette, transfer 10.0 cm³ to a conical flask.

5. Add about 20 cm^3 of distilled water, a few drops of indicator and titrate the mixture with 0.010 M HCl until the yellow solution just changes to red and remains red after shaking. (It may take a few moments for all the ammonia to transfer from the organic layer and react with the acid.

6. Titrate two more 10 cm^3 samples and complete a copy of Results Table 24a

7. Transfer the aqueous layer to a beaker. Rinse the pipette thoroughly, transfer 10.0 cm^3 to a flask, add about 20 cm^3 of water and a few drops of indicator solution, and titrate to the endpoint with 0.5 M HCl.

8. Titrate two more 10 cm^3 samples and complete a copy of Results Table 24b.

Results Table 24a

Solution in flask					mol dm^{-3}	cm^3
Solution in burette					mol dm^{-3}	
Indicator						
		Trial	1	2	3	4
Burette readings	Final					
	Initial					
Volume used/cm^3						
Mean titre/cm^3						

Results Table 24b

Solution in flask					mol dm^{-3}	cm^3
Solution in burette					mol dm^{-3}	
Indicator						
		Trial	1	2	3	4
Burette readings	Final					
	Initial					
Volume used/cm^3						
Mean titre/cm^3						

<u>Calculation</u>

1. Calculate the average concentration of ammonia in the organic layer from Results Table 24a.

2. Calculate the average concentration of ammonia in the aqueous layer from Results Table 24b.

3. Calculate the distribution coefficient.

$$K_D = \frac{[NH_3(aq)]}{[NH_3(tce)]}$$

EXPERIMENT 25
The pH of a weak acid at various concentrations

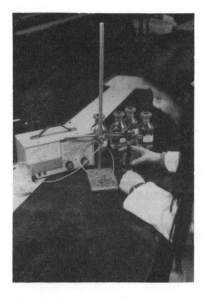

Aim

The purpose of this experiment is to examine the effect of dilution on the pH of ethanoic acid, a weak acid.

Introduction

Ethanoic acid dissociates according to the following equation:

$$CH_3CO_2H(aq) \rightleftharpoons CH_3CO_2^-(aq) + H^+(aq)$$

The extent of dissociation depends on the initial concentration of acid. By measuring the pH at different concentrations, you can see the effect of dilution. These results can be generalised for any weak acid.

Requirements

pH meter with glass electrode
wash-bottle of distilled water
a buffer solution (to calibrate the pH meter)
50 cm³ beaker
0.10 M, 0.010 M, 0.0010 M and 0.00010 M ethanoic acid solutions

Procedure

1. Calibrate the pH meter by dipping the glass electrode into a solution of known pH (a buffer solution) and turning the adjusting knob so that the scale shows the correct pH value. (If you are in doubt about this, ask your teacher.)

2. Rinse the glass electrode with distilled water and dip it into a beaker containing 0.00010 M ethanoic acid. Record the pH value in a copy of Results Table 25. Return the electrode to water; it must never be dry.

3. Rinse the beaker with the next solution, and repeat step 2, working from the most dilute solution to the most concentrated.

4. Calculate pH values for solutions of hydrochloric acid at the same concentrations and complete the final column of Results Table 25.

Results Table 25

Concentration of acid mol/dm^{-3}	Observed pH of solutions of ethanoic acid	Calculated pH of solutions of hydrochloric acid
0.00010	63	
0.0010		
0.010		
0.10		

Questions

1. Compare the pH of ethanoic acid with hydrochloric acid at each concentration.

    (a) In which of the two acids is the concentration of hydrogen ions greater?

    (b) What does this tell you about the extent of dissociation of ethanoic acid compared to hydrochloric acid?

2. (a) What happens to the difference between the pH of the two acids as concentration decreases? What does this tell you about the effect of dilution on dissociation?

    (b) Use Le Chatelier's principle to explain the effect of dilution on the extent of dissociation of ethanoic acid.

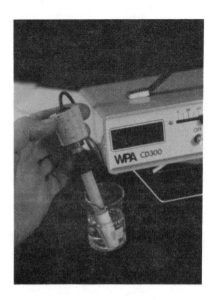

Aim

The purpose of this experiment is to compare
the strengths of acids by measuring pH at
the same concentration.

Introduction

You measure the pH of 0.010 M solutions of benzoic acid, $C_6H_5CO_2H$,
dihydrogenphosphate(V) ion, $H_2PO_4^-$, boric acid, H_3BO_3, and ethanoic acid,
CH_3CO_2H. From these measurements you can rank the acids in order of strength.

Requirements

pH meter with glass electrode
wash-bottle of distilled water
a buffer solution (to calibrate the pH meter)
beaker, 50 cm³
ethanoic acid solution, 0.010 M CH_3CO_2H
benzoic acid, 0.010 M $C_6H_5CO_2H$
boric acid solution, 0.010 M H_3BO_3
dihydrogenphosphate(V) ion solution, 0.010 M $H_2PO_4^-$

Procedure

Measure the pH of each solution, rinsing the electrode first each time, and
record the values in a copy of Results Table 26.

Results Table 26

Acid	Concentration /mol dm⁻³	pH
Boric acid	0.010	
Benzoic acid	0.010	
Ethanoic acid	0.010	
Dihydrogenphosphate(V) ion	0.010	

Question

List these acids in order of <u>decreasing</u> acid strength.

EXPERIMENT 27
The action of a buffer solution

Aim

The purpose of the experiment is to compare the effects of adding small amounts of acid and alkali to buffered and unbuffered solutions of the same pH.

Introduction

You are provided with a buffer solution designed to maintain a pH of 7.0 at 25 °C and some pure water which, if it is pure enough, should also have a pH of 7.0 at 25 °C.

To samples of these two liquids you add small measured amounts of 0.1 M NaOH and 0.1 M HCl, measuring the pH at each addition.

By comparing the pH changes in the two solutions you can demonstrate the action of a buffer solution.

Requirements

safety spectacles
2 burettes, 50 cm³, and stands
2 funnels, small
2 beakers, 100 cm³
hydrochloric acid, 0.1 M HCl
sodium hydroxide solution, 0.1 M NaOH ——————————————————————
measuring cylinder, 25 cm³
beaker, 50 cm³
distilled water - wash-bottle
buffer solution, pH 7.0
pH meter

pure water - fresh de-ionized water
  or freshly-boiled distilled water

(If no pH meter is available use:
  full-range pH paper, 1-14
  narrow-range pH paper
  thin glass stirring rod)

Procedure

1. Fill a burette with 0.1 M HCl and another with 0.1 M NaOH.

2. Using a measuring cylinder, put 25 cm³ of the buffer solution in a 50 cm³ beaker.

3. Rinse the pH meter electrode with distilled water from a wash-bottle, and put it into the beaker, making sure that the glass bulb is completely immersed. Set the meter to read 7.0.

4. Place the beaker under the burette containing NaOH and, making sure the alkali does not fall directly on to the electrode, add 1 drop of 0.1 M NaOH. Stir gently to ensure thorough mixing and record the pH in a copy of Results Table 27.

   If you cannot use a pH meter, measure the pH by transferring 1 drop of the mixed solution on a glass rod to a piece of pH paper. Use full-range paper first, and then narrow-range paper to obtain a more accurate value.

5. Add more NaOH to make the total volume added 1.0 cm³; measure and record the pH as before.

6. Add more NaOH to make the total volume added 5.0 cm³; measure and record the pH. Rinse the electrode in distilled water and stand it in a flask of distilled water.

7. Take another 25 cm³ portion of the buffer, and measure the pH on the addition of 1 drop, 1.0 cm³, and 5.0 cm³ of 0.1 M HCl in the same way as you did for NaOH. Again, rinse the electrode carefully and stand it in distilled water.

8. Put 25 cm³ of pure water in the beaker and, keeping its exposure time to the air as short as possible, measure its pH. If it is absolutely pure, its pH will be 7.0, but it is very difficult to achieve this. If the pH is less than 6.0, wash the beaker and electrode more carefully and try again.

9. When you have a pH between 6.0 and 7.0 for the 'pure' water, measure and record the pH changes on addition of 0.1 M NaOH and 0.1 M HCl just as you did for the buffer solution. Take special care to wash the electrode when you change from using alkali to acid. Record your results in Results Table 27.

10. Take another 25 cm³ of pure water, measure the pH and then leave it to stand open to the air for ten minutes. Measure the pH again and record the results.

Results Table 27

Volume added	pH on addition of 0.1 M NaOH to		pH on addition of 0.1 M HCl to	
	buffer	pure water	buffer	pure water
0				
1 drop				
1.0 cm³				
5.0 cm³				

pH of pure water with minimum air exposure	
pH of pure water after 10 mins air exposure	

Questions

1. By how much (to the nearest unit) does the pH of 25 cm³ of pure water change for the addition of 1.0 cm³ of 0.1 M HCl?

2. Calculate the ratio:

$$\frac{[H^+(aq)] \text{ after addition of 1.0 cm}^3 \text{ of HCl to pure water}}{[H^+(aq)] \text{ in pure water}}$$

3. By contrast, the addition of 1.0 cm³ 0.1 M HCl to 25 cm³ of the buffer solution should decrease the pH by about 0.1, and this corresponds to increasing $[H^+(aq)]$ by a factor of only about 1.25. However, the same number of hydrogen ions were added to both solutions. What must have happened to most of the hydrogen ions added to the buffer?

4. Why does the pH of pure water decrease when exposed to the air?

66

EXPERIMENT 28

Preparation of buffers: testing
their buffering capacity and the
effect of dilution

Aim

The purpose of this experiment is threefold:

1. to prepare two buffer solutions of
   pH values of 5.2 and 8.8;

2. to check the pH and buffering capacity
   of the prepared buffers;

3. to examine the effect of dilution of the
   prepared buffers on their pH and
   buffering capacity.

Introduction

This experiment lends itself to a planning exercise where you have to work
out for yourself the detailed procedure on the basis of the hints given below.
When you have worked out the details, check with your teacher that your
proposals are sensible and that apparatus will be available at a suitable
time. Try not to waste laboratory time on planning!

Hints

1. The following solutions will be available:

   ammonia solution, 1.0 M NH_3 ammonium chloride solution, 1.0 M NH_4Cl
   ethanoic acid, 1.0 M CH_3CO_2H sodium ethanoate solution, 1.0 M CH_3CO_2Na

2. Use the following equation to calculate how much of the above solutions
   you need to make 100 cm³ of each buffer:

   $$pH = pK_a - \log \frac{[HA(aq)]}{[A^-(aq)]}$$

   (You should be able to derive this from the expression for the dissociation constant, K_a, of a weak acid, HA.)

3. Your experience of Experiment 27 will be very useful.

4. For the dilution effect, try diluting your buffer to 1/10 and 1/100
   of its initial concentration.

5. Prepare a suitable table to record your results. This should be handed
   to your teacher with your answers to the questions below.

Questions

1. What did you use to make 100 cm³ of a buffer of pH 5.2?

2. What did you use to make 100 cm³ of a buffer of pH 8.8?

3. How close to the required values were the measured pH's of your prepared
   buffers?

4. Describe the effect of dilution on the pH of your prepared buffers. Was
   this effect expected?

5. Describe the effect of dilution on the buffering capacity of your prepared
   buffers.

6. Suggest a reason for the pH 8.8 buffer being less stable over a period of
   time than many other buffers.

EXPERIMENT 29
Determining the pH range
of some acid-base indicators

Aim

The purpose of this experiment is to show
that different acid-base indicators change
colour over different ranges of pH.

Introduction

To determine the pH range, you add an alkali, 1.0 cm³ at a time, to a
solution containing an indicator and a buffer, and measure the pH at each
addition. The pH values at which a colour change begins and ends define
the pH range.

It is best to measure the pH with a pH meter. However, if one is not available, you can calculate the pH from an equation we give at the end of the
experiment.

Requirements

safety spectacles
3 burettes, 50 cm³, and stands   ⎫
4 beakers, 250 cm³   ⎬   2 of these may be shared
3 funnels, small, for burettes   ⎭
distilled water
hydrochloric acid, 0.1 M HCl
sodium hydroxide solution, 0.1 M NaOH —————————————————————————————
pipette, 25 cm³, and safety filler
buffer solution, B
3 beakers, 100 cm³
indicator solutions, in dropping bottles, as follows:
   litmus, methyl orange, phenolphthalein, bromophenol blue, methyl red
pH meter (if available)

Procedure

1. Fill three burettes with distilled water, 0.1 M NaOH, and 0.1 M HCl
    respectively.

2. From burettes, run 25.0 cm³ of 0.1 M HCl and 10.0 cm³ of water into a
    small beaker. Similarly, run 25.0 cm³ of 0.1 M NaOH and 10.0 cm³ of
    water into a second small beaker.

3. Pipette 25.0 cm³ of solution B into a third beaker, add 2 drops of any
    one of the indicators provided, and mix. If the colour is very pale,
    add a few more drops of indicator, but always add the minimum amount
    which gives a distinct colour.

4. Add the same number of drops of indicator to the first two beakers as you
    did to the third. Put these two beakers aside to use for comparison
    purposes.

5. Add 0.1 M NaOH, 1.0 cm³ at a time, to the third beaker (containing B) and mix the contents thoroughly after each addition. At the <u>first sign</u> of a colour change, measure the pH of the mixture with a meter or, if a meter is not available, record the volume of alkali added. You will be able to detect a change in colour by comparing with the beaker containing 0.1 M HCl. Record the pH, or volume of alkali added, in a copy of Results Table 29a.

   To match colours precisely, you may find it helpful to dilute the 0.1 M HCl with water so that it has the same volume as the mixture of buffer solution and NaOH.

6. Continue adding 0.1 M NaOH, 1.0 cm³ at a time as before, until the colour change appears to be complete. Compare the solution with the beaker containing 0.1 M NaOH to help you to judge the end. Record the pH or the volume of added alkali at this point in Results Table 29a.

   To match colours precisely, you may find it helpful to dilute the 0.1 M NaOH with water so that it has the same volume as the mixture of buffer solution and added NaOH.

7. Repeat steps 1 to 6 for as many other indicators as time permits. If you have not measured the pH values, calculate them using the equation under Results Table 29a.

8. Complete a copy of Table 29b, using your experimental results and/or your data book. Choose suitable indicators so that their pH ranges, as shown by boxes similar to the two examples given, form a straight diagonal line from bottom left to top right.

Results Table 29a

Indicator used	Initial colour	Colour change starts		Colour change ends		Final colour
		Volume*	pH	Volume*	pH	

*Fill in these columns only if you have not measured the pH. Then calculate the pH from the volume, V cm³, of alkali added by using the equation:

$$pH = 3.0 + 0.23\ V \qquad (max\ V = 35)$$

Table 29b

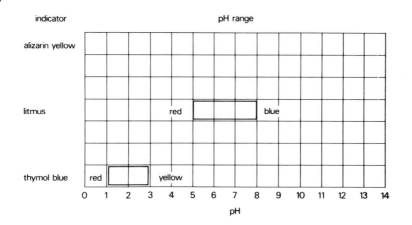

Question

Why was the indicator range measured in a buffer solution as opposed to a water solution?

EXPERIMENT 30
Determining the ionization
constant for an indicator

Aim

The purpose of this experiment is to
determine the ionization constant,
K_{In}, for bromophenol blue.

Introduction

There are two methods you can use. Both make use of the equation:

$$pH = pK_{In} - \log \frac{[HIn(aq)]}{[In^-(aq)]}$$

You prepare two solutions: one with a high hydrogen ion concentration in
which the indicator exists almost entirely as HIn; in the other solution,
with a high hydroxide ion concentration, the indicator exists almost entirely
as In^-.

These solutions are used to show the colour of the indicator at different
values of the ratio: $[HIn(aq)]/[In^-(aq)]$. From this measurement, you can
determine the value of K_{In}.

Ask your teacher which method to use, but it is a good idea to read the
following introductions for both methods.

Method 1 Prepare two sets of test-tubes in which the concentrations of the
two coloured forms of the indicator vary regularly, as shown in Fig. 24. You
can observe colours corresponding to different values of $[HIn(aq)]/[In^-(aq)]$
by looking through two tubes at once.

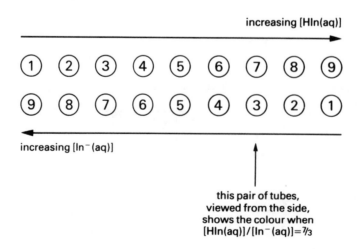

increasing [HIn(aq)]

increasing [In⁻(aq)]

this pair of tubes,
viewed from the side,
shows the colour when
[HIn(aq)]/[In⁻(aq)]=7/3

Fig. 24.

Method 2 Put two solutions containing the coloured forms in a piece of apparatus called a Bjerrum wedge. When viewed from the side, the varying widths of the two solutions give a continuous range of colours. Measurement of the widths at any point gives [HIn(aq)]/[In⁻(aq)], as in Fig. 25.

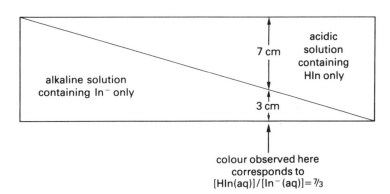

Fig. 25.

Requirements - Method 1

safety spectacles
measuring cylinder, 10 cm³
20 test-tubes
rack or racks to hold
  2 rows of 9 tubes each
bromophenol blue solution
2 teat pipettes
hydrochloric acid,
  concentrated, HCl — — — — — —
stirring rod
sodium hydroxide
  solution, 4 M NaOH — — — — — — -
distilled water
buffer solution, pH 3.7

Requirements - Method 2

safety spectacles
2 beakers, 250 cm³
hydrochloric acid,
  0.1 M HCl
measuring cylinder, 25 cm³
bromophenol blue solution
Bjerrum wedge
stirring rod
sodium hydroxide solution,
  0.1 M NaOH
beaker, 100 cm³
buffer solution, pH 3.7

Procedure - Method 1

1. Pour about 5 cm³ of bromophenol blue solution into a test-tube and add 1 drop of concentrated hydrochloric acid. In this solution, almost all of the indicator is in the form HIn. Call this solution 'X'.

2. Pour about 5 cm³ of bromophenol blue solution into a test-tube and add 1 drop of 4 M sodium hydroxide solution. In this solution, almost all of the indicator is in the form In⁻. Call this solution 'Y'.

3. Arrange 18 test-tubes in a rack (or racks) as shown in Fig. 24 and carefully add drops of solutions X and Y corresponding to the numbers in Fig. 24.

4. Add 10 cm³ distilled water to each tube and stir if necessary so that the colour is even from top to bottom.

5. Pour 10 cm³ of the buffer solution (pH = 3.7) into a test-tube and add 10 drops of bromophenol blue solution. Shake the tube to mix the contents.

6. By holding the buffer tube alongside the pairs of tubes in the rack, find the pair which is nearest to the same colour.

    You will find it helpful to view the tubes against a brightly lit white background.

7. Record the value of [HIn(aq)]/[In⁻(aq)] in the pair of tubes which best matches the buffer solution.

Procedure - Method 2

1. To 250 cm³ 0.1 M HCl in a beaker, add 10 cm³ of bromophenol blue solution. Stir well, and then pour the solution into one half of the Bjerrum wedge. In this solution almost all of the indicator is in the form HIn.

2. To 250 cm³ 0.1 M NaOH in a beaker, add 10 cm³ of bromophenol blue solution. Stir well, and then pour the solution into the other half of the wedge. In this solution almost all of the indicator is in the form In⁻.

3. To 25 cm³ of the buffer solution in a small beaker, add 1.0 cm³ of bromophenol blue solution. Measure the volume of indicator carefully so that its concentration is the same in all three beakers. Stir well, and then pour the solution into the cell which fits on top of the wedge.

4. Slide the cell along the wedge until the colour in the cell is the same as the colour in the wedge immediately beneath it, as in Fig. 26. You will find it helpful to view the wedge against a brightly lit white background.

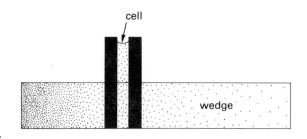

Fig.26.

5. Record the value of [HIn(aq)]/[In⁻(aq)] corresponding to the colour match. This can be obtained by direct measurement, as in Fig. 25. Difficulty in colour matching may be due to different concentrations; did you measure the volumes of indicator carefully enough?

Calculation - both methods

Substitute your measured value of [HIn(aq)]/[In⁻(aq)] and the given pH of the buffer in the equation:

$$\text{pH} = \text{p}K_{\text{In}} - \log \frac{[\text{HIn(aq)}]}{[\text{In}^-\text{(aq)}]}$$

and obtain a value for $\text{p}K_{\text{In}}$. Obtain K_{In} from $\text{p}K_{\text{In}}$ in the same way as you calculated hydrogen ion concentration from pH. Compare your result with the value given in your data book.

Questions

1. Give two advantages and two disadvantages of the wedge method compared with the test-tube method. You should be able to answer without having done both methods - look at the introduction.

2. Why was as much as 10 drops (Method 1) or 1 cm³ (Method 2) of indicator solution added to the buffer solution?

3. Could a buffer of pH 8 have been used in this experiment?

EXPERIMENT 31

Determining the dissociation constant
of a weak acid, using an indicator

Aim

The purpose of the experiment is to prepare
a buffer solution from benzoic acid and
sodium benzoate, and then to determine its
pH by a visual method, as used in Experiment
30. pK_a for the acid can then be calculated
from the pH by the usual buffer formula.

Introduction

First you prepare a buffer solution from benzoic acid and sodium benzoate.
Then you measure its pH using the same procedure as in Experiment 30.
Having determined the pH, you calculate K_a for benzoic acid using the
equation:

$$pH = pK_a - \log \frac{[C_6H_5CO_2H(aq)]}{[C_6H_5CO_2^-(aq)]}$$

Ask your teacher which of the two methods you should use, but again read the
introduction for both methods given for Experiment 30.

Requirements - Method 1

safety spectacles
benzoic acid solution, 0.020 M $C_6H_5CO_2H$
sodium benzoate solution, 0.020 M $C_6H_5CO_2Na$
measuring cylinder, 10 cm³
20 test-tubes
rack or racks to hold 2 rows of 9 tubes each
bromophenol blue solution
2 teat pipettes
hydrochloric acid, concentrated, HCl — — — — — — — — — — — — — — — — — —
stirring rod
sodium hydroxide solution, 4 M NaOH — — — — — — — — — — — — — — — — —
distilled water

Requirements - Method 2

safety spectacles
benzoic acid solution, 0.020 M $C_6H_5CO_2H$
sodium benzoate solution, 0.020 M $C_6H_5CO_2Na$
measuring cylinder, 25 cm³
beaker, 100 cm³
2 beakers, 250 cm³
hydrochloric acid, 0.1 M HCl
bromophenol blue solution
Bjerrum wedge
stirring rod
sodium hydroxide solution, 0.1 M NaOH

Procedure - Method 1

1. Prepare a buffer solution by mixing 5 cm³ of 0.020 M benzoic acid and 5 cm³ of 0.020 M sodium benzoate in a test-tube.

2. Follow the procedure for Method 1 of Experiment 30. In step 5, use the buffer you have just prepared.

Procedure - Method 2

1. Prepare a buffer solution by mixing 12.5 cm³ of 0.020 M benzoic acid and 12.5 cm³ of 0.020 M sodium benzoate in a small beaker.

2. Follow the procedure for Method 2 of Experiment 30. In step 3, use the buffer you have just prepared.

Calculation - both methods

1. Obtain the pH of the buffer solution by substituting in the equation:

$$pH = pK_{In} - \log \frac{[HIn(aq)]}{[In^-(aq)]}$$

2. Obtain pK_a, and hence K_a, by substituting in the equation:

$$pH = pK_a - \log \frac{[C_6H_5CO_2H(aq)]}{[C_6H_5CO_2^-(aq)]}$$

EXPERIMENT 32
Obtaining pH curves for
acid-alkali titrations

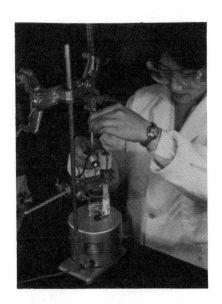

Aim

The purpose of this experiment is to obtain
curves which show how the pH changes during an
acid-base titration and to use these curves
to choose suitable indicators for different
combinations of acid and alkali.

Introduction

You have available two acids, one strong, one weak, and two alkalis, one
strong, one weak, giving four possible combinations. You should study the
results of each of the four combinations even if you cannot complete four
yourself.

The four types of pH curve
correspond to four types
of titration using a strong
or weak acid and a strong or
weak base.

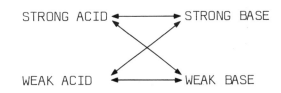

For each combination, you run the alkali from a burette, in small steps, into
25 cm³ of the acid and measure the pH at each addition. You then plot pH
against volume of alkali added. It is best to measure pH with a pH meter,
but it is possible to get adequate results using narrow-range pH paper if a
meter is not available. Using a magnetic stirrer speeds up the experiment
but it is not essential.

Requirements

safety spectacles
pH meter or narrow-range pH papers to cover pH 1 to 14
ethanoic acid, 0.100 M CH₃CO₂H
pipette, 25 cm³, and safety filler
beaker, 100 cm³ or conical flask, 250 cm³ and thin glass rod
magnetic stirrer (if available)
2 burettes, 50 cm³, and stands
2 funnels, small, for filling burettes
4 beakers, 250 cm³, with labels
ammonia solution, 0.100 M NH₃
distilled water
hydrochloric acid, 0.100 M HCl
sodium hydroxide solution, 0.100 M NaOH -------------------------

1. Pipette 25.0 cm^3 of 0.100 M ethanoic acid into a 100 cm^3 beaker, add a small stirring paddle, and stand the beaker on a magnetic stirrer.

2. Carefully clamp the electrode of a pH meter so that the bulb is completely immersed in the acid and is clear of the stirring paddle. If a magnetic stirrer (or mechanical stirrer) is not available, you will have to swirl the beaker by hand. If a pH meter is not available, it is better to use a conical flask rather than a beaker.

3. Fill a burette with 0.100 M ammonia solution and clamp it over the beaker (or flask) containing the acid.

4. Obtain a value of the pH of the acid, either by reading the pH meter (which must, of course, have been calibrated) or by removing <u>1 small drop</u> on a thin glass rod and testing with narrow-range indicator paper. If you do not know which range to use, test first with full-range Universal indicator paper. However, it is important not to remove more than the minimum acid from the flask.

5. Add the alkali from the burette in steps as shown in Results Table 32 and record the pH at each addition after thorough mixing.

    (The volumes shown in the tables were chosen on the assumption that the acids and alkalis have <u>precisely</u> the same concentration, so that the equivalence point is at $\overline{25.0 \text{ cm}^3}$ added alkali. If this is not the case, the volumes will have to be adjusted somewhat. Ask your teacher's advice on this.)

6. Repeat steps 1-5 with at least one other combination of acid and alkali, and complete a copy of Results Table 32. If you do not have time to do all the experiments, you can share data with other students, or use our specimen results.

7. Plot four curves, on <u>one</u> sheet of graph paper. The usual convention is to show the pH on the vertical scale, and the volume of added alkali on the horizontal scale.

Questions (Answer for <u>each</u> pH curve)

1. Over what section of the curve (if any) did the pH change by 2 units for the addition of 0.2 cm^3 (or less) of alkali? Answer in the form of 'pH y to pH z'.

2. At what volume of alkali was the equivalence point?

3. By how much (approximately) would the pH change on the addition of one extra drop of alkali (0.05 cm^3) at 12.5 cm^3, 25.0 cm^3, 34.0 cm^3 of added alkali?

4. What indicator(s) (if any) would be suitable for this titration? To be suitable, the indicator must have an end-point which coincides with the equivalence point, <u>and</u> the end-point must be 'sharp', i.e. the colour change must be comp<u>le</u>te for a very small addition of alkali - preferably one drop (0.05 cm^3).

Results Table 32

0.100 M NH₃ added to 25.0 cm³ of 0.100 M CH₃CO₂H										
Volume/cm³	0.0	5.0	10.0	15.0	20.0	24.0	24.2	24.4	24.6	24.8	24.9
pH										
Volume/cm³	24.95	25.0	25.05	25.1	25.2	25.4	25.6	25.8	26.0	30.0	35.0
pH										

0.100 M NaOH added to 25.0 cm³ of 0.100 M CH₃CO₂H										
Volume/cm³	0.0	5.0	10.0	15.0	20.0	24.0	24.2	24.4	24.6	24.8	24.9
pH										
Volume/cm³	24.95	25.0	25.05	25.1	25.2	25.4	25.6	25.8	26.0	30.0	35.0
pH										

0.100 M NH₃ added to 25.0 cm³ of 0.100 M HCl										
Volume/cm³	0.0	5.0	10.0	15.0	20.0	24.0	24.2	24.4	24.6	24.8	24.9
pH										
Volume/cm³	24.95	25.0	25.05	25.1	25.2	25.4	25.6	25.8	26.0	30.0	35.0
pH										

0.100 M NaOH added to 25.0 cm³ of 0.100 M HCl										
Volume/cm³	0.0	5.0	10.0	15.0	20.0	24.0	24.2	24.4	24.6	24.8	24.9
pH										
Volume/cm³	24.95	25.0	25.05	25.1	25.2	25.4	25.6	25.8	26.0	30.0	35.0
pH										

EXPERIMENT 33
The variation of boiling-point with
composition for a binary liquid mixture

Aim

The purpose of this experiment is to
construct boiling-point/composition
curves for mixtures of two different
liquids.

Introduction

In this experiment you study one or more of the following three systems:

1. trichloromethane and methyl ethanoate;

2. ethanol and cyclohexane;

3. propan-1-ol and propan-2-ol.

You measure the boiling-points of these mixtures at various compositions and
construct boiling-point/composition curves from your results. Inversion of
these curves into vapour pressure/composition curves will then enable you to
describe the deviation from ideality (if any) for each mixture. For simplicity,
you plot volume composition rather than mole fraction - the curves have the
same general shape.

Requirements

safety spectacles
ground-glass-joint apparatus in Fig. 27 or Fig. 28
thermometer, 0-100 °C, ± 0.1 °C
anti-bumping granules
2 burettes (dry) and stands
2 small funnels
test-tube
Bunsen burner, gauze and tripod
trichloromethane, $CHCl_3$ —
methyl ethanoate, $CH_3CO_2CH_3$ ⎫
ethanol, C_2H_5OH ⎪
cyclohexane, C_6H_{12} ⎬ — — — — — — — — — — — — — — — — —
propan-1-ol, $CH_3CH_2CH_2OH$ ⎪
propan-2-ol, $CH_3CH(OH)CH_3$ ⎭

Hazard warning

Methyl ethanoate, ethanol, cyclohexane, propan-1-ol and propan-2-ol
are all flammable. Trichloromethane is toxic.

Therefore you MUST:

KEEP THE STOPPERS ON THE BOTTLES AS MUCH AS POSSIBLE

KEEP THE BOTTLES AWAY FROM FLAMES

USE TRICHLOROMETHANE IN THE FUME CUPBOARD

WEAR SAFETY SPECTACLES

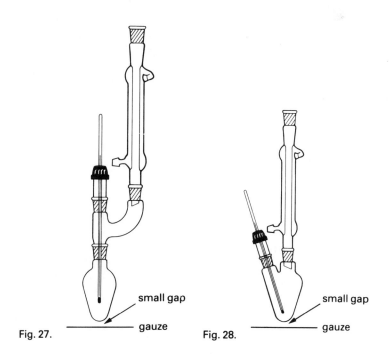

Fig. 27.   small gap   gauze   Fig. 28.   small gap   gauze

Procedure

1. Set up a suitable assembly for reflux, either Fig. 27 or Fig. 28, with the flask positioned over a tripod and gauze. It is important that the thermometer is positioned so that it will dip into the liquid mixture but it must not touch the walls of the flask.

2. Choose one of the three systems shown in Table 33a and, if possible, ensure that the other students in your class investigate the other systems. In the remaining procedure steps we refer to the components of each system as A or B, as shown in Table 33a.

Table 33a

System	Component A	Component B
1	trichloromethane, $CHCl_3$	methyl ethanoate, $CH_3CO_2CH_3$
2	ethanol, C_2H_5OH	cyclohexane, C_6H_{12}
3	propan-1-ol, $CH_3(CH_2)_2OH$	propan-2-ol, $CH_3CH(OH)CH_3$

3. Pour 20 cm³ of A and 20 cm³ of B into separate labelled burettes.

4. Transfer 10.0 cm³ of A from the burette to the pear-shaped flask containing a few anti-bumping granules. Heat the flask gently until the liquid just begins to boil.

5. Record the boiling-point of A in a copy of Results Table 33b.

6. Turn off the Bunsen burner and allow the apparatus to cool for about two minutes.

7. Measure 2.0 cm³ of liquid B from the burette into a test-tube and pour this liquid down the condenser into the pear-shaped flask.

8. Reheat the flask gently until the liquid mixture just boils and record its boiling-point.

9. Repeat stages 6, 7 and 8 with further additions of 2 cm³ portions of component B until a total of 10 cm³ of B has been added.

10. Allow the apparatus to cool and ask your teacher how you can safely dispose of the mixture.

11. Repeat the experiment from step 4, this time starting with 10 cm³ of
    liquid B in the flask and adding 2 cm³ of component A after each boiling-
    point determination until a total of 8 cm³ of A has been added.

Results Table 33b

Volume/cm³		% composition	Boiling-point/°C		
A	B	A (by volume)	System 1	System 2	System 3
10	0	100			
10	2	83.3			
10	4				
10	6				
10	8				
10	10				
0	10				
2	10				
4	10				
6	10				
8	10				

Treatment of results

1. Plot a graph of boiling-point (y-axis) against percentage composition by
    volume (x-axis). Label the graph as shown in Fig. 29, substituting the
    names of your liquids for A and B.

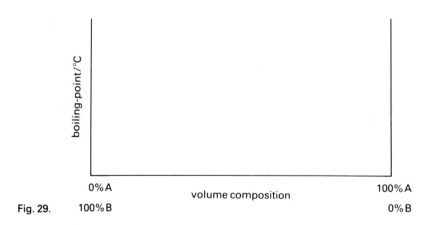

Fig. 29.

2. Collect the results for the other two systems from other students and
    plot the results in exactly the same way.

Questions

1. For each boiling-point/composition curve, deduce an approximate vapour
    pressure/composition curve. Insert dotted lines to represent the
    vapour pressures you would expect for 'ideal' mixtures.

2. Study the vapour pressure curves and classify each mixture as ideal or
    as showing positive or negative deviations from ideality.

EXPERIMENT 34
Measuring the temperature change
on forming solutions

Aim

The purpose of this experiment is to
measure the change in temperature which
occurs on mixing two components of a
non-ideal mixture and to see how it
correlates with the vapour pressure/
composition curves obtained in
Experiment 33.

Introduction

The enthalpy change on mixing two miscible liquids A and B, $\Delta H^{\ominus}_{mix}$, can be
defined as the heat change which occurs when 1 mole of liquid A is mixed with
1 mole of liquid B under standard conditions. However, we are not particularly concerned with the numerical value of $\Delta H^{\ominus}_{mix}$ but mainly with its sign.
This is easy to establish from the temperature change on mixing liquids A
and B. An increase in temperature indicates that the sign is negative
whereas a decrease in temperature indicates that the sign is positive.
Using the same systems as in Experiment 33, you determine the sign of $\Delta H^{\ominus}_{mix}$
for each pair of liquids. This gives information about the intermolecular
forces acting between A ... A, B ... B and A ... B.

Requirements

safety spectacles
boiling-tube
cotton wool
beaker, tall form, 250 cm³
2 measuring cylinders, 10 cm³
2 small funnels
thermometer, -5 to 50 °C, ± 0.1 °C
trichloromethane, CHCl₃ —
methyl ethanoate, CH₃CO₂CH₃ ⎫
ethanol, C₂H₅OH ⎪
cyclohexane, C₆H₁₂ ⎬ — — — — — — — — — — — — — — — — — —
propan-1-ol, CH₃CH₂CH₂OH ⎪
propan-2-ol, CH₃CH(OH)CH₃ ⎭

Procedure

1. Decide which of the three mixtures you are going to use and, if possible, ensure that other students in your class pick the other mixtures.

2. Place a boiling-tube into a beaker and surround the tube with cotton wool, as shown in Fig. 30.

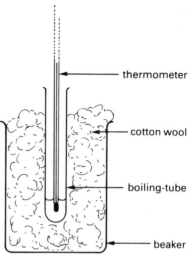

Fig. 30.

3. Pour 10 cm³ of one of the components (liquid A) from a measuring cylinder into the insulated boiling-tube.

4. Measure 10 cm³ of liquid B in a clean measuring cylinder.

5. Record the temperature of liquid A, T_A, and liquid B, T_B. Record the average value as T_1 in a copy of Results Table 34.

6. Pour the liquid from the measuring cylinder into the liquid in the insulated tube and stir carefully with the thermometer. Record the new temperature, T_2, and the temperature difference, ΔT.

7. Dispose of the mixture according to your teacher's instructions.

Results Table 34

Mixture	$T_A/°C$	$T_B/°C$	$T_1/°C$	$T_2/°C$	$\Delta T/°C$
1. Trichloromethane and methyl ethanoate					
2. Ethanol and cyclohexane					
3. Propan-1-ol and propan-2-ol					

Questions

1. State the sign of $\Delta H^{\ominus}_{mix}$ for each mixture.

2. Which of the three mixtures most closely approaches ideal behaviour? Explain your answer.

3. Explain the observed temperature change on mixing trichloromethane and methyl ethanoate in terms of the interactions between the molecules.

4. Explain the shape of the vapour pressure curve for this system as determined in Experiment **33**, based on your answer to 3.

5. Correlate the vapour pressure/composition curve with the sign of ΔH_{mix} for the ethanol/cyclohexane system.

EXPERIMENT 35

Determining the approximate strength
of a hydrogen bond

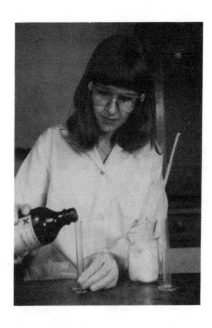

Aim

The purpose of this experiment is to
determine the strength of the hydrogen
bonds formed between trichloromethane
and methyl ethanoate. You can achieve
this by measuring the enthalpy change
on mixing, ΔH_{mix}.

Introduction

Because this is a planning experiment, we give fewer details and instructions
than you have been used to. It is similar to Experiment 34 but you must
consider carefully the quantities of reagent to be used. You should also
look up the specific heat capacities of the reagents or ask your teacher.

Requirements

Make a list of requirements including the amounts needed; show the list to
your teacher or technician.

Hazard warning

Methyl ethanoate is flammable and trichloromethane is toxic.

Therefore you MUST:

PERFORM THIS EXPERIMENT IN A FUME CUPBOARD

KEEP THE STOPPER ON THE BOTTLE AS MUCH AS POSSIBLE

WEAR GLOVES AND SPECTACLES

Procedure

Work this out for yourself and keep an accurate record. Hint: consider the
possibility of an equilibrium reaction and the effect of using an excess of
one reagent.

Results and calculation

Tabulate your results clearly, showing in your calculations how you
arrived at the strength of the hydrogen bond (kJ mol^{-1}).

Discuss your results with your teacher and fellow students.

EXPERIMENT 36

The effect of hydrogen bonding on liquid flow

Aim

The purpose of this experiment is to compare the viscosities of four different liquids and to interpret the results in terms of hydrogen bonding.

Introduction

You are supplied with sealed tubes containing four different liquids. Each tube has a small air bubble trapped at one end. If the tubes are inverted in turn, the time it takes for the air bubble to travel through the length of the tube can be taken as a measure of the intermolecular forces in the liquid.

Requirements

4 sealed tubes containing:
  propan-1-ol
  propane-1,2-diol
  propane-1,2,3-triol
  propane-1,2,3-triyl triethanoate
stopclock or stopwatch

Procedure

Invert each tube in turn and measure the time it takes for the air bubble to travel through the length of the tube. Record your results in a copy of Results Table 36.

Results Table 36

Liquid	Formula	Time/s
Propan-1-ol	$CH_3CH_2CH_2OH$	
Propane-1,2-diol	$CH_3CH(OH)CH_2OH$	
Propane-1,2,3-triol	$CH_2(OH)CH(OH)CH_2OH$	
Propane-1,2,3-triyl triethanoate	$CH_2(OCOCH_3)CH(OCOCH_3)CH_2OCOCH_3$	

Questions

1. Account for the differences in viscosity between the three alcohols you have investigated.

2. Propane-1,2,3-triyl triethanoate has much larger molecules than propane-1,2,3-triol, and yet it is much less viscous. Why is this?

Aim

The purpose of this experiment is to test
nine liquids in order to find out if their
molecules are polar.

Introduction

In this experiment you study the effect
of a charged rod on a stream of liquid
from a burette. A deflection of the
stream indicates that the liquid consists of polar molecules.

The extent of the deflection, under
standard conditions, shows how polar
the molecules are.

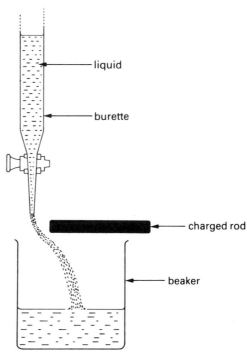

Fig. 31.

Requirements

9 dry beakers, with labels
9 labelled burettes in stands, each one corked and containing one of the
    following liquids:
    cyclohexane, C_6H_{12}
    cyclohexene, C_6H_{10}
    distilled water, H_2O
    hexane, C_6H_{14}
    hexan-1-ol, $C_6H_{13}OH$
    methyl ethanoate, $CH_3CO_2CH_3$
    propanone, CH_3COCH_3
    tetrachloromethane, CCl_4
    trichloromethane, $CHCl_3$
9 small funnels
polythene rod
piece of fur or suitable cloth

Procedure

1. Rub the polythene rod with the fur. This will give it a negative charge.

2. Position a beaker beneath the jet of one of the burettes as shown in Fig. 31. Remove the cork and allow a stream of the liquid to run from the burette with the tap fully open.

3. Bring the charged rod close to the stream of liquid and note any deflection that occurs. Also note the extent of deflection on an arbitrary scale from 0 to 3 (0 = no deflection; 3 = greatest deflection). Record your results in a copy of Results Table 37.

4. Pour the liquid from the beaker back into the labelled burette to avoid waste. Leave this set up in a fume cupboard with the cork replaced in the top of the burette. This is then ready for other students to use.

5. Repeat the above procedure with each liquid in turn. Try to standardize the conditions, otherwise your rating of the deflection will be worthless. The liquid level in the burettes, the tap aperture, the position of the rod, and the extent to which the rod is charged should always be the same.

Results Table 37

Compound	Formula	Structural formula	Deflection
Cyclohexane			
Cyclohexene			
Hexane			
Hexan-1-ol			
Methyl ethanoate			
Propanone			
Tetrachloromethane			
Trichloromethane			
Water			

Questions

1. Explain the effect of the charged rod on a jet of water.

2. What do you think would happen with a rod of opposite charge? Explain.

3. Place the liquids in an approximate order of decreasing polarity. Interpret this order in terms of the structure of each molecule and comment particularly on the different results in the following pairs.

    (a) trichloromethane, $CHCl_3$, and tetrachloromethane, CCl_4;

    (b) cyclohexane, C_6H_{12}, and cyclohexene, C_6H_{10}.

EXPERIMENT 38
Determining relative molecular mass
by a freezing-point method

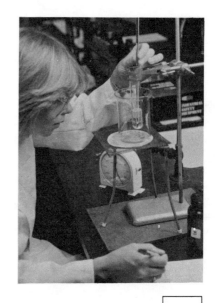

Aim

The purpose of this experiment is
to give you an opportunity to test
your practical skills. The
experiment also serves as an example
of a freezing-point method being
used to calculate relative molecular
mass.

Introduction

The procedure below is taken from an A-level practical examination
paper; read it carefully and report fully, as directed.

Requirements

safety spectacles
compound A (about 40 g)
compound B (about 3 g)
2 dry boiling-tubes and stoppers
access to balance capable of weighing to 0.01 g
2 beakers, 400 cm³
Bunsen burner, tripod, gauze and mat
retort stand, boss and clamp
thermometer 0-100 °C (in 1 °C)
stop-clock

Procedure

You are provided with a compound, A, which melts at 53.0 °C, and an organic
compound, B.

You are required to determine the relative molecular mass of compound B,
using a freezing-point method.

Previous experience of this type of experiment is not required; full details
of the method and calculation are given below.

Experiment 1

(a) Weigh a clean, dry boiling-tube empty and then containing between 0.70 g
and 0.80 g of B. Now add between 16.50 g and 18.00 g of A and re-weigh.
Record the masses in (a copy of) the table provided.

(b) Place a 400 cm³ beaker containing water on a tripod and gauze, and warm
to about 70 °C. Set the boiling-tube containing A and B vertically in
a clamp, dipping into the water, and loosely stopper. Allow the
contents to melt; when molten remove the stopper and stir gently with a
clean, dry thermometer to ensure thorough mixing.

(c) When the liquid is completely mixed, remove the water bath and replace
it with an empty beaker. Stir the mixture gently and continuously.
Start the stop-clock when the temperature reaches 57 °C ± 0.5 °C.
Measure the temperatures, estimating to the nearest quarter of a degree,
at half-minute intervals. stirring throughout, and record them in
(a copy of) the table provided.

Experiment 2

Repeat Experiment 1, using a second boiling-tube and fresh materials, but weighing between 1.45 g and 1.55 g of B. Record your results as before in the table provided.

Results Table 38a

	Experiment 1	Experiment 2
Mass of boiling-tube + B/g		
Mass of boiling-tube/g		
Mass of B/g		
Mass of boiling-tube + A + B/g		
Mass of A/g		

Results Table 38b Experiment 1

Time/min	0	$\frac{1}{2}$	1	$1\frac{1}{2}$	2	$2\frac{1}{2}$	3	$3\frac{1}{2}$	4	$4\frac{1}{2}$	5	$5\frac{1}{2}$	6	$6\frac{1}{2}$	7	$7\frac{1}{2}$	8
Temp./°C																	

Results Table 38c Experiment 2

Time/min	0	$\frac{1}{2}$	1	$1\frac{1}{2}$	2	$2\frac{1}{2}$	3	$3\frac{1}{2}$	4	$4\frac{1}{2}$	5	$5\frac{1}{2}$	6	$6\frac{1}{2}$	7	$7\frac{1}{2}$	8
Temp./°C																	

Plot a temperature/time curve for each experiment, selecting suitable scales for the vertical temperature axis and the horizontal time axis.

Distinguish the two graphs by circumscribing points from the first experiment with circles, i.e. $\odot$ and indicating those from the second experiment by crosses, i.e., ×. Using the graphs deduce freezing-points, t_1 and t_2 respectively, of each mixture.

Calculation

The relative molecular mass, M, of compound B may be shown to be given approximately by the expression:

$$M = \frac{7.10 \times \text{mass of } B \text{ (g)} \times 10^3}{[53.0 - t(°C)] \times \text{mass of } A \text{ (g)}}$$

where t is the melting-point of the mixture.

Use this expression to calculate a value for M from each of the two experiments.

Suppose that a mass of B equal to the combined masses used in both experiments has been dissolved in a mass of A equal to that used in the first experiment. At what temperature would you expect this mixture to melt?

EXPERIMENT 39

Determining enthalpy changes and
volume changes of solution

Aim

The purpose of this experiment is to measure
the enthalpy changes on forming solutions of
several salts and to determine the accompany-
ing volume changes.

Introduction

In Experiment 10, you determined the enthalpy change of solution for the
process:

$$NH_4Cl(s) + 100H_2O(l) \rightarrow NH_4Cl(aq, 100 \ H_2O)$$

In part A of this experiment you determine, in the same way, the enthalpy
change of solution for several ionic compounds. This time, however, you work
with 2 M solutions. In part B you measure the volume change which occurs when
each of the solutions is formed and relate the enthalpy change of solution
to the percentage volume change.

If you use a burette to record the volume change in the solution, you will
need to take account of the volume of liquid contained in the uncalibrated
portion (or 'dead space'), below the 50 cm^3 mark of the burette. This
procedure is not necessary if you use a gas-measuring tube.

Requirements

safety spectacles
disposable gloves
2 burettes and stoppers to fit (or 1 gas-measuring tube)
2 retort stands, bosses and clamps
wash-bottle of distilled water
polystyrene cup
thermometer, 0-100 °C (in 0.5 °C)
spatula
weighing bottle with tight-fitting cap or stopper
lithium chloride, LiCl, anhydrous ⎫
sodium chloride, NaCl, anhydrous ⎪
potassium chloride, KCl, anhydrous ⎬ (finely powdered)
calcium chloride, CaCl$_2$, anhydrous ⎪
iron(III) chloride, FeCl$_3$, anhydrous ⎭ — — — — — — — — — — — — —
access to balance weighing to 0.01 g

Procedure

A. Enthalpy of solution, ΔH_{soln}

1. Measure 50.0 cm^3 of distilled water from a burette into a polystyrene
    cup. Note the initial temperature of the water and record it in a
    copy of Results Table 39b.

2. Weigh out 0.1 mol of a salt (see Table 39a for molar mass values) as quickly as possible, in a weighing bottle, to prevent the salt absorbing moisture from the air. (Note that all salts absorb moisture from the atmosphere - some more readily than others.)

   Record the masses in a copy of Results Table 39b.

3. Dissolve the salt in the distilled water by careful stirring with the thermometer. Note the highest or lowest temperature of the solution and record it in a copy of Results Table 39b.

4. Calculate the temperature change, ΔT, of the liquid and, thus, ΔH_{soln}. Record your answers in a copy of Results Table 39b.

B. Volume change

1. Transfer 50.0 cm^3 of distilled water from one burette to another. Note the reading on the second burette, V_2, and record it in a copy of Results Table 39c.

2. Weigh out 0.1 mol of the salt observing the precautions given in part A and record the mass in a copy of Results Table 39c.

3. Add the salt to the water in the second burette, stopper the burette and shake it carefully to dissolve the salt. (For $CaCl_2$ and $FeCl_3$, the solid should be added in small portions and the burette cooled under the tap. Do not insert the bung until the temperature rise is complete.)

4. Read the final volume, V_3, when the solution has reached room temperature.

5. From the expression given in Results Table 39c, calculate the % volume change. In order to calculate the volume of the solid, V_1, you will need to know the densities of the solids shown in Table 39a.

Table 39a

Salt	Density/g cm^{-3}	Molar mass/g mol^{-1}
LiCl	2.1	42.4
NaCl	2.2	58.4
KCl	2.0	74.6
CaCl$_2$	2.5	111
FeCl$_3$	2.8	162

Results Table 39b

Salt	Mass/g			Temperature/^{o}C			ΔH_{soln}/kJ mol^{-1} = total mass (kg) x ($-\Delta T$) x 10 x 4.18 kJ kg^{-1} K^{-1}
	Empty bottle	Bottle + salt	Salt	Initial	Final	Change	
LiCl							
NaCl							
KCl							
CaCl$_2$							
FeCl$_3$							

Results Table 39c

| Salt | Mass of salt/g | Volume of salt, V_1/cm^3 | Burette reading | | % volume change, $\dfrac{V_3 - V_2 - V_1}{50 + V_1} \times 100$ |
			Initial, V_2/cm^3	Final, V_3/cm^3	
LiCl					
NaCl					
KCl					
CaCl$_2$					
FeCl$_3$					

Questions

1. Explain why a factor of 10 is introduced into the equation for ΔH_{soln} given in Results Table 39b.

2. (a) What trend is observed in ΔH_{soln} for the Group I chlorides?

   (b) How does this relate to the size of the cation?

3. (a) What trend is observed in ΔH_{soln} for the series KCl, CaCl$_2$, and FeCl$_3$?

   (b) How does this relate to the charge on the cation?

4. In terms of the interactions between particles in solute, solvent and solution, explain the observed enthalpy change, ΔH_{soln}, for one solid. Take CaCl$_2$ as an example.

5. (a) Explain how the volume reduction occurs.

   (b) What is the correlation between volume change and enthalpy of solution ΔH_{soln}? Suggest a reason.

EXPERIMENT 40

Investigating the hydrolysis of
benzenediazonium chloride

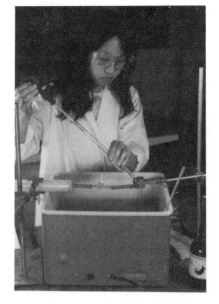

Aim

The purpose of this experiment is to determine the rate equation for the reaction in
which benzenediazonium chloride is hydrolysed
and hence find the order of reaction with
respect to benzenediazonium chloride.

Introduction

Benzenediazonium chloride is an unstable substance, which decomposes when
heated above 5 °C to give phenol, nitrogen and hydrochloric acid:

$$C_6H_5N_2{}^+Cl^-(aq) + H_2O(l) \rightarrow C_6H_5OH(aq) + N_2(g) + HCl(aq)$$

There are several stages in the preparation of the reaction mixture, so you
do this at a temperature low enough for the rate of reaction to be negligible..
When you are ready, you warm the mixture quickly to a fixed temperature and
measure the volume (V_t) of gas produced at one minute intervals for about
25 minutes. You then leave the mixture until no further reaction appears to
be occurring, and measure the total volume (V_∞) of gas produced since the
clock was started.

Because the volume of gas produced in time t is proportional to the amount
of benzenediazonium chloride used up, it follows that:

$$V_\infty \quad \propto \quad [C_6H_5N_2{}^+Cl^-(aq)] \text{ at the start}$$

$$V_\infty - V_t \quad \propto \quad [C_6H_5N_2{}^+Cl^-(aq)] \text{ at time } t$$

A plot of $(V_\infty - V_t)$ against time will therefore have the same form as the
concentration/time graph and can be used to obtain information about the
rate of reaction.

Note that, in this experiment, you need not attempt to judge the time when
the reaction begins. The clock can be started at any time after the mixture
has reached a steady temperature. Even if some benzenediazonium chloride
has reacted by then, the amount remaining can be taken as giving the
'initial' concentration, and this can be calculated, if necessary, from V_∞.

safety spectacles and protective gloves
water-bath, thermostatically controlled, set between 40 and 50 °C
thermometer, 0-100 °C
side-arm test-tube with bung
3-way tap
glass syringe, 100 cm³
rubber tubing (2 short lengths)
2 retort stands, bosses and clamps
wash-bottle of distilled water
measuring cylinder, 10 cm³
test-tube
sodium nitrite, $NaNO_2$ —
spatula
beaker 250 cm³
crushed ice
hydrochloric acid, concentrated, HCl — — — — — — — — — — — — — —
pumice or anti-bumping granules
graduated pipette, 5 cm³ or 10 cm³
pipette filler
phenylamine, $C_6H_5NH_2$ — — — — — — — — — — — — — — — — — — —
teat-pipette
stopclock or stopwatch

Hazard warning

Phenylamine is toxic, by ingestion and by skin absorption.
It is also flammable and gives off a harmful vapour.

Concentrated hydrochloric acid is corrosive and gives off a
harmful vapour. Therefore you MUST:

WEAR SAFETY SPECTACLES AND GLOVES
WORK IN A FUME CUPBOARD WHERE POSSIBLE
KEEP BOTTLES AWAY FROM FLAMES
KEEP STOPPERS ON BOTTLES AS MUCH AS POSSIBLE

Procedure

1. Set the control on the water bath to a temperature between 40 and 50 °C
    and hang a thermometer in it. If possible this should be done before the
    lesson so that it has time to reach a steady temperature.

2. Set up the apparatus, without the chemicals, as shown in Fig. 32.

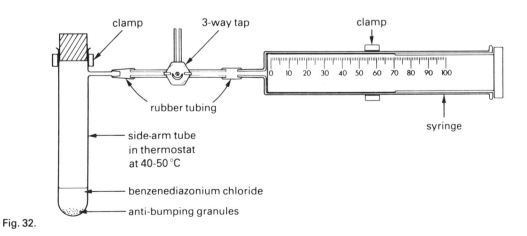

Fig. 32.

3. When you have connected the three-way tap between the side-arm tube and
    the syringe, turn the tap so that it connects the syringe with the open
    air as in Fig. 33A. Press the plunger in, so that the syringe is empty,
    and then turn the tap so that it connects the side-arm tube with the air
    as in Fig. 33B.

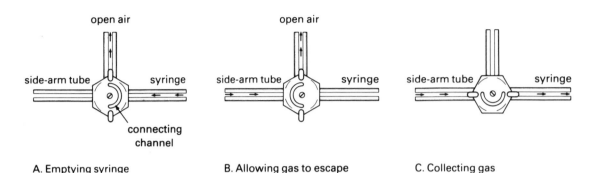

Fig. 33.

4. Using a 10 cm³ measuring cylinder, add 2.0 cm³ of distilled water into
    a test-tube. Weigh 0.80 g of sodium nitrite and dissolve it in the
    water. Cool this solution in a beaker of crushed ice and keep it handy.

5. Remove the side-arm tube and, using a measuring cylinder, measure
    into it 5.0 cm³ of distilled water and 2.5 cm³ of concentrated
    hydrochloric acid.

6. Add a few anti-bumping or pumice granules to the side-arm tube and seal
    its mouth with a rubber bung.

7. Place the side-arm tube in a beaker of crushed ice and, using a
    pipette and filler, add 1.0 cm³ of phenylamine. Mix the contents
    of the tube thoroughly and allow them to cool.

8. Using a teat-pipette, add the sodium nitrite solution a few drops at a
    time to the phenylamine solution in the side-arm tube. Shake the tube
    gently to swirl its contents as you do so.

9. Check that the thermostat temperature has become constant (at about
    45 °C), then clamp the side-arm tube in position in the thermostat
    (see Fig. 32) so that the solution is completely immersed in water.

10. Wait four minutes with the tap open to the air, so that the solution
    warms to the temperature of the water-bath. Record the temperature.

11. After four minutes, turn the tap so that it connects the side-arm tube
    with the syringe (Fig. 33C), reset the clock and take a reading.
    Regard this as time zero.

12. Continue taking readings each minute for about 25 minutes. Enter your
    results in a larger copy of Results Table 40a. The total volume of gas
    produced is over 100 cm³, so as you see the volume approaching the
    100 cm³ mark, get ready to turn the tap to the position shown in Fig. 33A
    (emptying the syringe into the air). Just as it reaches 100 cm³, turn the
    tap, expel the collected nitrogen, then turn the tap back to the position
    shown in Fig. 33C (connecting the side-arm tube with the syringe).
    Depending on how you make up your mixture, you may have to do this twice
    during the experiment.

13. After 25 minutes, leave the apparatus for at least another half hour for the reaction to finish completely. This gives you the total volume of nitrogen produced in the reaction, V_∞ (i.e., the volume produced at infinite time). If you want to speed things up, immerse the side-arm tube into a beaker of hot water at about 60 °C. If the syringe is nearly full, empty it first, noting the volume of nitrogen expelled. The volume of nitrogen should reach its maximum in a few minutes. Remember to allow the syringe to cool back down to the temperature of the water-bath before taking your final reading (V_∞).

Results Table 40a

Time, t/min								
Volume of N_2, V_t/cm³								
$(V_\infty - V_t)$/cm³								

Calculations

1. Work out the values of $(V_\infty - V_t)$ and enter them into your copy of Results Table 40a.

2. Plot $(V_\infty - V_t)$ (vertical axis) against time (horizontal axis). Draw a smooth curve through the points.

3. Construct tangents to your curve and measure the slope at each point. Draw one at time 0 and at least four others evenly spaced.

4. Use your graph to complete a larger copy of Results Table 40b.

Results Table 40b

Time /min	Slope /cm³ min⁻¹	Rate /cm³ min⁻¹	$(V_\infty - V_t)$/cm³

5. Plot another graph of rate of reaction against $(V_\infty - V_t)$, which is proportional to the concentration of benzenediazonium chloride.

Questions

1. Use the information from your second graph to write a rate equation for the reaction.

2. Use your second graph to work out a value for the rate constant, k, for the reaction, including its units.

3. Explain why the escape of gas during the first four minutes, before recording the first volume reading, can be neglected.

EXPERIMENT 41

The kinetics of the reaction between
iodine and propanone in aqueous solution

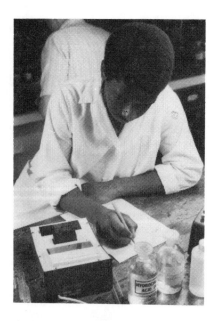

Aim

The purpose of this experiment is to
obtain the rate equation for the reaction
between iodine and propanone by determining the order of reaction with respect
to each reactant and to the catalyst
(hydrogen ions). The equation is:

I_2(aq) + CH_3COCH_3(aq) →

$\qquad$ CH_3COCH_2I(aq) + H^+(aq) + I^-(aq)

Introduction

A catalyst does not necessarily appear in the stoichiometric equation (here
it appears as a product) but it can appear in the rate equation. The other
species which are likely to appear are the reactants and so you may assume
that the rate equation is:

$$\text{rate} = k[CH_3COCH_3]^p[I_2]^q[H^+]^r$$

You will be determining the order of reaction with respect to each reactant
by varying the concentration of each species in turn, keeping the others
constant and following the reaction colorimetrically. As the intensity of
the iodine colour decreases more light is transmitted through the solution
(i.e. the absorbance decreases).

There are three parts to the experiment:

1. Choosing the right filter for the colorimeter.

2. Calibrating the colorimeter so that meter readings can be converted to
    concentrations of iodine.

3. Obtaining values for the concentration of iodine at intervals of time
    for a series of experiments with the following sets of conditions:

    (a) Initial $[CH_3COCH_3]$ varying; $[I_2]$, $[H^+]$ constant

    (b) Initial $[I_2]$ varying; $[CH_3COCH_3]$, $[H^+]$ constant

    (c) Initial $[H^+]$ varying; $[CH_3COCH_3]$, $[I_2]$ constant

    Each set of experiments gives you the order of reaction with respect to
    one component. If there are three groups of students in your class
    working on this experiment, then we suggest that each group assumes
    responsibility for one set of conditions. Sharing your results will
    speed up matters considerably.

Requirements

safety spectacles
colorimeter with a set of filters
set of optically matched test-tubes to fit colorimeter (with stoppers)
wash-bottle of distilled water
4 burettes with stands, filling funnels and beakers
iodine solution, 0.020 M I_2 (in KI(aq))
propanone solution, 2.0 M CH_3COCH_3
hydrochloric acid, 2.0 M HCl
stopclock or watch
thermometer, 0 to 100 °C

Procedure

1. Choose a filter. First, switch on the colorimeter to allow it to warm up. (Leave it switched on until you have finished, unless your teacher advises otherwise.) Ideally, the filter should let through only light of the particular wavelength absorbed by the coloured solution. So, for a red solution which absorbs cyan light, you would use a cyan filter - cyan is the complementary colour to red (see Fig. 34. Since iodine solution is reddish, use a filter in the blue-green range. To select the best, proceed as follows.

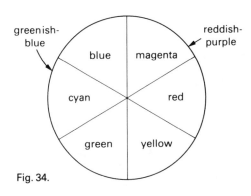

Fig. 34.

   (a) Put any one of the suitable filters into the slot in the colorimeter and insert a tube of distilled water, covering it to exclude stray light. Turn the adjusting knob so that the meter shows zero absorbance (or, on some colorimeters, 100% transmission). Mark the rim of the tube so that you can replace it in the same position.

   (b) Replace the tube with a tube containing the most concentrated iodine solution you will use (see Results Table 41a) and take a reading of absorbance (or % transmission). Mark this tube too so that you can replace it in the same position.

   (c) Repeat steps (a) and (b) above for other suitable filters in turn.

   (d) Choose the filter which gives the greatest absorbance (or least transmission).

2. Prepare a calibration curve, so that you can convert your meter readings taken during the experiment to concentration of iodine.

   (a) To do this, prepare a series of iodine solutions as suggested in Results Table 41a. Measure and record the absorbance of each one. Most colorimeters are liable to 'drift', so you should zero the machine before each reading, by inserting the distilled water 'blank' and re-setting the needle to 100% transmission.

Results Table 41a

Volume of 0.020 M I_2 solution/cm^3	Volume of distilled water/cm^3	$[I_2(aq)]$ /mol dm^{-3}	Meter reading (% absorbance or transmission)
0.0	10.0	0	
1.0	9.0	0.0020	
2.0	8.0	0.0050	
3.0	7.0	0.0060	
4.0	6.0	0.0080	
5.0	5.0	0.010	

   (b) Plot a graph of meter reading against concentration of iodine and keep this to use in analysing your results.

3. Decide which series of experiments you will do from Table 41b below and note their letters. The figures in heavy type will help you choose. If no other students are working on this investigation then you will need to do all the experiments.

Table 41b

	Experiment						
	a	b	c	d	e	f	g
Volume of 2 M propanone/cm^3	2	4	6	2	2	2	2
Volume of 0.02 M iodine/cm^3	2	2	2	4	1	2	2
Volume of 2 M HCl /cm^3	2	2	2	2	2	4	6
Volume of water /cm^3	4	2	0	2	5	2	0
[Propanone] /mol dm^{-3}	**0.4**	**0.8**	**1.2**	0.4	0.4	0.4	0.4
[I$_2$]/mol dm^{-3}	**0.004**	0.004	0.004	**0.008**	**0.002**	0.004	0.004
[H$^+$]/mol dm^{-3}	**0.4**	0.4	0.4	0.4	0.4	**0.8**	**1.2**

4. For your first mixture, measure out iodine solution, acid and water from burettes into a test-tube. Wipe the tube clean and handle only at the top.

5. Measure the propanone solution into another test-tube, again using a burette. Keep the outside of all tubes clean and dry.

6. Adjust the colorimeter to zero against the tube of distilled water.

7. Add the propanone solution to the first mixture and start the clock. Quickly stopper the test-tube and invert it six or seven times to mix the contents thoroughly.

8. Put the tube in the colorimeter in time to take a reading at 30 seconds after the clock was started.

9. Take further readings at 30 second intervals for 6 minutes, or until the colour disappears, if this occurs sooner. Record your results in a copy of Results Table 41c.

10. Record the temperature of the room and the temperature of the mixture after the final reading. If they differ by more than 2 or 3 °C you may be advised to repeat the experiment, modifying the procedure as in step 12.

11. Repeat steps 4 to 10 for the other two mixtures in your set of experiments.

12. If time permits, repeat your measurements. Some colorimeters may give better results if you remove the tube after each reading, and reset the meter to zero just before re-inserting the reaction tube in time for the next reading.

Results Table 41c

Time/min	0	$\frac{1}{2}$	1	$1\frac{1}{2}$	2	$2\frac{1}{2}$	3	$3\frac{1}{2}$	4	$4\frac{1}{2}$	5	$5\frac{1}{2}$	6
Meter reading													
$[I_2(aq)]/10^{-3}$ mol dm^{-3}													

Treatment of results

1. Use your calibration curve to convert the meter readings to iodine concentrations and enter these values in Results Table 41c. Calculate a value of initial concentration from the data in Table 41b.

2. Plot graphs of concentration of iodine (vertical axis) against time (horizontal axis) for each mixture in the set of experiments you have done. Plot all three on the same sheet of graph paper.

3. From these graphs obtain values for the initial rate of reaction. Note that, under the conditions used in this experiment, the graphs will probably be straight lines. In this case the initial rate is simply the slope of the line. If the graphs are curves, simply draw a tangent to the curve at time zero and measure its slope.

4. Determine the order of reaction with respect to the component you have been varying, by comparing the initial rates at different concentrations.

Questions

1. Study the graphs plotted by other groups of students for varying the concentrations of the other two components. Work out the order of reaction with respect to each of these components by comparing the initial rates at different concentrations. Use all the information you have collected to write the rate equation for the reaction.

2. What is the overall order of the reaction?

3. Compare the rate equation with the stoichiometric equation for the reaction. What is the main difference between the two? How do you explain this difference?

4. Calculate the rate constant for the overall reaction using each of the three initial rates from your set of experiments and average the results. Compare this with values obtained from the other sets of experiments.

5. What are the main sources of error in this experiment?

EXPERIMENT 42
Determining the activation energy
of a reaction

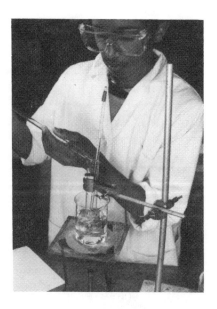

Aim

The purpose of this experiment is to
determine the activation energy, E_a,
for the reduction of peroxo-
disulphate(VI) ions, $S_2O_8{}^{2-}$(aq), by
iodide ions, I^-(aq), using a
'clock' reaction.

Introduction

The equation for the reduction of peroxodisulphate(VI) ions by iodide ions is:

$$S_2O_8{}^{2-}(aq) + 2I^-(aq) \rightarrow 2SO_4{}^{2-}(aq) + I_2(aq)$$

A small, known amount of thiosulphate ions is added to the reaction mixture,
which also contains some starch indicator. The thiosulphate reacts with the
iodine formed in the above reaction as in the following equation:

$$2S_2O_3{}^{2-}(aq) + I_2(aq) \rightarrow S_4O_6{}^{2-}(aq) + 2I^-(aq)$$

At the instant that all the thiosulphate has reacted, free iodine is
produced in the solution and its presence is shown by the appearance of the
blue-black colour of the iodine-starch complex, i.e. the thiosulphate ions
act as a 'monitor' indicating the point at which a certain amount of iodine
has been formed. For this reason the reaction is often referred to as an
iodine 'clock' reaction. In general, for a 'clock' reaction:

Rate of reaction $\propto 1/t$ where t is the time taken to reach a specified stage.

You carry out the experiment at five different temperatures between about
20 °C and 50 °C. You then find the activation energy for the reaction by
plotting a graph of log $(1/t)$ against $1/T$ (T is the absolute temperature)

Requirements

safety spectacles
beaker, 400 cm³
2 thermometers, 0-100 °C
Bunsen burner, tripod, gauze and mat
4 burettes and stands, with beakers and funnels for filling
2 boiling-tubes
clamp and stand
potassium peroxodisulphate(VI) solution, 0.020 M $K_2S_2O_8$
potassium iodide solution, 0.50 M KI
sodium thiosulphate solution, 0.010 M $Na_2S_2O_3$
starch solution, 0.2%
stopclock or watch

Procedure

1. Half-fill the beaker with water and heat it to between 49 °C and 51 °C. This will be used as a water-bath.

2. Using a burette, measure out 10 cm³ of potassium peroxodisulphate(VI) solution into the first boiling-tube. Clamp this in the water-bath and place a thermometer in the solution in the boiling-tube.

3. Using burettes, measure out 5 cm³ each of the potassium iodide and sodium thiosulphate solutions and 2.5 cm³ of starch solution into the second boiling-tube. Place another thermometer in this solution and stand it in the water-bath.

4. When the temperatures of the two solutions are equal and constant (to within ± 1 °C), pour the contents of the second boiling-tube into the first, shake to mix, and start the clock.

5. When the blue colour of the starch-iodine complex appears, stop the clock and write down the time in a copy of Results Table 42.

6. Repeat the experiment at temperatures close to 45 °C, 40 °C, 35 °C, 30 °C. (The temperatures you use may differ from those by a few degrees but must, of course, be recorded carefully.)

Results Table 42

Temperature/°C					
Temperature, T/K					
Time, t/s .					
$\log_{10}(1/t)$					
$\frac{1}{T}$/K^{-1}					

Calculations

1. Plot a graph of $\log_{10}(1/t)$ (vertical axis) against $1/T$ (horizontal axis).

2. Use your graph to calculate a value for the activation energy.

EXPERIMENT 43

Determining the activation energy
of a catalysed reaction

Aim

The purpose of this experiment is to
determine the activation energy for the
oxidation of iodide ions by peroxo-
disulphate(VI) ions in the presence of
iron(III) ions. You then compare the
value obtained with that for the
uncatalysed reaction, determined in
Experiment 42.

Introduction

Because this is a planning experiment, we give fewer details and instructions
than you have been used to. It is, of course, very similar to Experiment 42
but you should consider carefully which concentrations of solutions to use
since the reaction is catalysed. You should also consider the effect of
temperature on reaction rate.

Requirements

Make a list of requirements including the masses and amounts needed: discuss
the list with your teacher or technician at least a day before you want to do
the experiment.

Procedure

Work this out for yourself and keep an accurate record.

Results

1. Tabulate your results in an appropriate form.

2. Calculate a value for the activation energy of this reaction.

Question

Compare your result with the activation energy you calculated in
Experiment 42. Comment on the two values, and discuss them with your teacher.

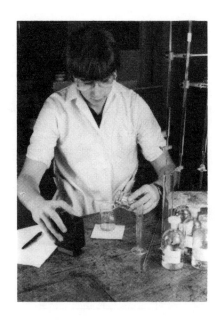

Aim

The purpose of this experiment is to
determine the rate equation for the
reaction between bromide and bromate(V)
ions in aqueous solution.

Introduction

Bromide and bromate(V) ions in acid solution react according to the equation:

$$5Br^-(aq) + BrO_3^-(aq) + 6H^+(aq) \rightarrow 3Br_2(aq) + 3H_2O(l) \dots\dots\dots\dots\dots(1)$$

In order to follow the reaction, two other substances are added to the
reaction mixture.

(a) A precisely known, small amount of phenol. This reacts immediately
   with the bromine produced, removing it from solution:

$$3Br_2(aq) + C_6H_5OH(aq) \rightarrow C_6H_2Br_3OH(aq) + 3H^+(aq) + 3Br^-(aq) \dots\dots(2)$$

(b) Methyl orange solution, which is bleached colourless by free bromine:

$$Br_2(aq) + \text{methyl orange} \rightarrow \text{bleached methyl orange} \dots\dots\dots\dots(3)$$
$$\text{(acid form: pink)} \qquad \text{(colourless)}$$

As soon as all the phenol has reacted with bromine produced in
reaction (1), free bromine will appear in solution and bleach the methyl
orange. If the time taken for the methyl orange solution to be
bleached is t, then the rate of reaction (1) is proportional to $1/t$.

In the experiment you study the effect on the rate of reaction (1) of varying
the concentration of bromide ions, bromate (V) ions, and hydrogen ions in turn,
with the concentrations of the others held constant. To save time, you could
cooperate with other students and pool your results.

Requirements

safety spectacles
phenol solution, 0.00010 M C_6H_5OH
wash-bottle of distilled water
3 burettes and stands, with beakers and funnels for filling
measuring cylinder, 25 cm^3
potassium bromide solution, 0.010 M KBr
potassium bromate(V) solution, 0.0050 M $KBrO_3$
acidified methyl orange solution, labelled C, 0.001%
white tile
thermometer 0-100 °C
2 beakers, 100 cm^3
stopclock or watch
sulphuric acid, 0.010 M H_2SO_4
potassium bromate(V) solution, 0.20 M $KBrO_3$
methyl orange solution, labelled D, 0.001% in 0.40 M KBr

Procedure

A. Varying the concentration of bromide ions

1. Prepare the first pair of mixtures in two beakers, as specified in Table 44a. Use burettes to measure the potassium bromide and phenol solutions and the water; use a measuring cylinder for the others.

Table 44a

Beaker X		Beaker Y		
Volume of 0.01 M KBr /cm³	Volume of H_2O /cm³	Volume of 0.005 M $KBrO_3$/cm³	Volume of solution C /cm³	Volume of 0.00010 M phenol/cm³
10.0	0	10.0	15.0	5.0
8.0	2.0	10.0	15.0	5.0
6.0	4.0	10.0	15.0	5.0
5.0	5.0	10.0	15.0	5.0
4.0	6.0	10.0	15.0	5.0
3.0	7.0	10.0	15.0	5.0

2. Have ready a copy of Results Table 44b.

Results Table 44b

Volume of $Br^-(aq)$/cm³	10.0	8.0	6.0	5.0	4.0	3.0
Time, t/s						
$\frac{1}{t}$/10^{-2} s^{-1}						
Temperature/°C						

Average temperature of solutions =   °C

3. Pour the contents of beaker X into beaker Y and start the clock. Mix the solutions by pouring from one beaker to the other, twice, and place the beaker containing the mixture on the white tile.

4. When the pink colour disappears, stop the clock and record the time in a copy of Results Table 44b. (As you are looking for a disappearance of colour, this may need a little practice. If in doubt, repeat the reaction once or twice until you get consistent times.) Record the temperature of the solution.

5. Work through the rest of the mixtures in Table 44a in the same way, recording each result as you go.

B. Varying the concentration of bromate(V) ions

6. Follow a similar procedure to that for part A, but keep the volume of bromide solution constant at 10.0 cm³ and vary the volume of bromate(V) solution as shown in Table 44c.

Table 44c

Beaker X		Beaker Y		
Volume of 0.005 M KBrO$_3$/cm^3	Volume of H$_2$O /cm^3	Volume of 0.01 M KBr /cm^3	Volume of solution C /cm^3	Volume of 0.00010 M phenol/cm^3
10.0	0	10.0	15.0	5.0
8.0	2.0	10.0	15.0	5.0
6.0	4.0	10.0	15.0	5.0
5.0	5.0	10.0	15.0	5.0
4.0	6.0	10.0	15.0	5.0
3.0	7.0	10.0	15.0	5.0

Record your results in a copy of Results Table 44d.

Results Table 44d

Volume of BrO$_3{}^-$(aq)/cm^3	10.0	8.0	6.0	5.0	4.0	3.0
Time, t/s						
$\frac{1}{t}$/10^{-2} s^{-1}						
Temperature/°C						

Average temperature of solutions =   °C

C. Varying the concentration of hydrogen ion

7. For this part of the experiment you need different solutions, as stated in Table 44e. Follow a similar procedure to that for part A.

Table 44e

Beaker X			Beaker Y	
Volume of 0.01 M H$_2$SO$_4$/cm^3	Volume of H$_2$O /cm^3	Volume of 0.20 M KBrO$_3$/cm^3	Volume of solution D /cm^3	Volume of 0.00010 M phenol/cm^3
10.0	0	10.0	15.0	5.0
8.0	2.0	10.0	15.0	5.0
6.0	4.0	10.0	15.0	5.0
5.0	5.0	10.0	15.0	5.0
4.0	6.0	10.0	15.0	5.0
3.0	7.0	10.0	15.0	5.0

Record your results in a copy of Results Table 44f.

Results Table 44f

Volume of acid/cm^3	10.0	8.0	6.0	5.0	4.0	3.0
Time, t/s						
$\frac{1}{t}$/10^{-2} s^{-1}						
Temperature/°C						

Average temperature of solutions =   °C

Treatment of results

1. For each part of the experiment, plot a graph of $1/t$ against volume of the reactant under consideration. $1/t$ is proportional to the rate of reaction and the volume of reactant is proportional to the concentration, since the total volume is constant.

2. Deduce from each graph whether or not the reaction is first order with respect to the reactant under consideration.

3. If you think the reaction is not first order, plot another graph, as explained below.

    Suppose that, for a reactant A, rate = $k_1[A]^n$

    Under the conditions of this experiment, it follows that

    $1/t = k_2 V^n$ (V is the initial volume of reactant)

    $\therefore$ $\log(1/t)$ = $n\log V + \log k_2$

    Plotting $\log(1/t)$ against $\log V$ should therefore give a straight line with slope n.

Questions

1. Write the rate equation for the reaction.

2. Qualitatively compare the rates of the three reactions stated in the introduction.

3. Why does the phenol solution need to be very dilute?

EXPERIMENT 45
Some simple redox reactions

Aim

The purpose of this experiment is to illustrate
some redox chemistry by means of simple test-tube
reactions between metals and salt solutions.

Introduction

You place zinc, copper and silver into salt solutions containing the ions
Zn^{2+}, Cu^{2+} and Ag^+. The observations you record will provide a starting
point for your study of redox reactions.

Requirements

safety spectacles
6 test-tubes in a rack
2 strips of copper foil
2 strips of zinc foil
2 pieces of silver wire
emery paper
copper sulphate solution, 0.5 M $CuSO_4$
zinc sulphate solution, 0.5 M $ZnSO_4$
silver nitrate solution, 0.1 M $AgNO_3$
silver residues bottle

Procedure

1. If necessary, clean the strips of metal with emery paper.
2. Dip a piece of each metal into solutions of salts of each of the other
   two metals. Leave for about two minutes and examine each test-tube,
   recording your observations in a copy of Results Table 45.

Results Table 45

Solution	Observations		
	Zn	Cu	Ag
Zn^{2+}(aq)			
Cu^{2+}(aq)			
Ag^+(aq)			

Questions

1. Write ionic equations for the reactions you have observed.
2. In each case, state which of the reactants has been oxidized and which
   has been reduced. Show any changes in oxidation number.
3. Which of the metals acts as both reducing agent and oxidizing agent?

EXPERIMENT 46
Measuring the potential difference generated by some simple electrochemical cells

Aim

The purpose of this experiment is to construct three electrochemical cells, and to measure the potential difference between the electrodes, noting the polarity.

Introduction

You combine the following half-cells and measure the potential difference between the metal electrodes using a high resistance voltmeter.

Table 46a

Cell	Half-cells	
1	$Zn^{2+}(aq) + 2e^- \rightleftharpoons Zn(s)$	$Cu^{2+}(aq) + 2e^- \rightleftharpoons Cu(s)$
2	$Ag^+(aq) + e^- \rightleftharpoons Ag(s)$	$Cu^{2+}(aq) + 2e^- \rightleftharpoons Cu(s)$
3	$Ag^+(aq) + e^- \rightleftharpoons Ag(s)$	$Zn^{2+}(aq) + 2e^- \rightleftharpoons Zn(s)$

When you measure the potential difference between the electrodes of each cell, the polarity of each electrode is indicated by the positive sign and negative sign on the voltmeter (or red and black terminals respectively). Thus, you can determine at which electrode electrons are needed and at which electrode electrons are produced. From this, you can work out the direction in which the overall reaction proceeds.

Requirements

safety spectacles
strip of copper foil
strip of zinc foil
silver wire
emery paper, 3 small pieces
4 beakers, 50 cm³
3 connecting leads with crocodile clips attached
copper sulphate solution, 1.0 M CuSO₄
zinc sulphate solution, 1.0 M ZnSO₄
silver nitrate solution, 0.10 M AgNO₃
filter paper strips
potassium nitrate solution, saturated, 3 M KNO₃
voltmeter, high resistance

Procedure

1. If necessary, clean each metal strip (or wire) with a separate piece of emery paper.

2. Place each metal strip in a separate beaker. Hold each strip vertically against the inside of the beaker so that about 2 cm projects above the rim. Fold the projection down over the rim of the beaker and clamp it in position with a crocodile clip attached to a lead.

3. Pour about 20 cm³ of the appropriate salt solution into each beaker so that each metal strip dips into a solution of its own ions. Make sure the crocodile clips keep dry.

4. Prepare a salt bridge by soaking a strip of filter paper in saturated potassium nitrate solution. Let the surplus solution drain off by hanging the strip over the fourth beaker.

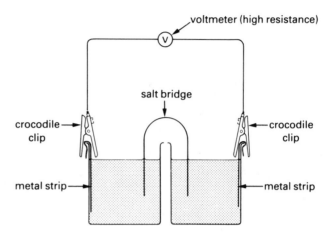

Fig. 35.

5. Connect the zinc and copper electrodes to a high-resistance voltmeter, as shown in Fig. 35, and complete the circuit with the salt bridge.

6. If the reading on the voltmeter is negative, reverse the connections to obtain a positive reading. Record the potential difference in a copy of Results Table 46b, noting which electrode is positive (connected to the red terminal of the voltmeter) and which is negative.

7. Remove the salt bridge as soon as possible and throw it away. Disconnect the voltmeter.

8. Repeat steps 4 to 7 for the other two cells.

Results Table 46b

Cell	Positive electrode	Negative electrode	Potential difference/V
1			
2			
3			

Questions

1. Consider cell 1:

    (a) What reaction is taking place at the zinc electrode to make it negative?

    (b) What reaction is taking place at the copper electrode to make it positive?

2. Similarly for the other cells, write the half-reactions which make the negative electrode negative and the positive electrode positive.

3. Add each pair of half-reactions to give the overall cell reaction.

4. Compare these cell reactions to the test-tube reactions in Experiment 45.

5. What is the function of the salt bridge in the cells?

EXPERIMENT 47
Testing predictions about redox reactions

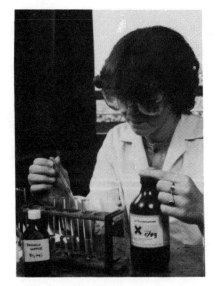

Aim

The purpose of this experiment is to test whether the use of $E^\ominus$ values to predict the outcome of redox reactions is reliable.

Introduction

You are to mix the reactants underlined in equations A to F below and by simple observation or simple tests decide whether a reaction has taken place. You can then compare your results with predictions based on $E^\ominus$ values for the appropriate half-cells.

A. $\underline{Br_2(aq) + 2I^-(aq)} \rightarrow 2Br^-(aq) + I_2(aq)$

B. $\underline{Br_2(aq) + 2Cl^-(aq)} \rightarrow 2Br^-(aq) + Cl_2(aq)$

C. $\underline{Zn(s) + 2Fe^{3+}(aq)} \rightarrow Zn^{2+}(aq) + 2Fe^{2+}(aq)$

D. $\underline{2MnO_4^-(aq) + 16H^+(aq) + 10Br^-(aq)} \rightarrow 8H_2O(l) + 2Mn^{2+}(aq) + 5Br_2(aq)$

E. $\underline{2MnO_4^-(aq) + 16H^+(aq) + 5Cu(s)} \rightarrow 8H_2O(l) + 2Mn^{2+}(aq) + 5Cu^{2+}(aq)$

F. $\underline{3S_2O_8^{2-}(aq) + 2Cr^{3+}(aq) + 7H_2O(l)} \rightarrow 6SO_4^{2-}(aq) + Cr_2O_7^{2-}(aq) + 14H^+(aq)$

Requirements

safety spectacles
10 test-tubes in rack, with corks or bungs to fit
potassium iodide solution, 0.1 M KI
bromine water, $Br_2(aq)$ —
1,1,1-trichloroethane, CH_3CCl_3 —
potassium chloride solution, 0.1 M KCl
iron(III) chloride solution, 0.1 M $FeCl_3$
spatula
zinc powder, Zn
potassium hexacyanoferrate(III)(ferricyanide) solution, 0.1 M $K_3Fe(CN)_6$ —
potassium manganate(VII) (permanganate) solution, 0.02 M $KMnO_4$
sulphuric acid, dilute, 1 M H_2SO_4
potassium bromide solution, 0.1 M KBr
copper powder, Cu
potassium peroxodisulphate (persulphate) solution, 0.1 M $K_2S_2O_8$
chromium(III) chloride solution, 0.1 M $CrCl_3$

Procedure

A. Reaction between $Br_2(aq)$ and $I^-(aq)$

1. Place about 3 cm^3 of potassium iodide solution in a test-tube and add, dropwise, about the same volume of bromine water.

2. Cork and shake the tube and note any change.

3. Add a little 1,1,1-trichloroethane, shake and observe the colour of the bottom layer. Record your observations and deductions in a larger copy of Results Table 47.

Results Table 47

Reaction	Additional test	Observations	Deductions	Predictions from $E^\ominus$
A				
B	▨			
C				
D				
E	▨			
F	▨			

B. Reaction between $Br_2(aq)$ and $Cl^-(aq)$

1. Place about 3 cm^3 of potassium chloride solution in a test-tube and add about the same volume of bromine water.

2. Cork and shake the tube. Record your observations and deductions.

C. Reaction between $Fe^{3+}(aq)$ and $Zn(s)$

1. Place about 3 cm^3 of iron(III) chloride solution in a test-tube and add a very small amount of zinc powder (a mere pinch).

2. Cork and shake the tube and allow any solid to settle.

3. Test for the presence of $Fe^{2+}(aq)$ by adding a few drops of potassium hexacyanoferrate(III) solution ($Fe^{2+}(aq)$ will give a dark blue precipitate if present). Record your observations and deductions.

D. Reaction between $MnO_4^-(aq)$, $H^+(aq)$ and $Br^-(aq)$

1. Place about 3 cm^3 of potassium manganate(VII) solution in a test-tube and add about 1 cm^3 of dilute sulphuric acid. Mix the solutions.

2. Place about 3 cm^3 of potassium bromide solution in another test-tube and add, dropwise, the mixture from step 1.

3. Cork and shake the tube and note any colour change.

4. Add a little 1,1,1-trichloroethane, shake and observe the colour of the bottom layer.

E. Reaction between $MnO_4^-(aq)$, $H^+(aq)$ and $Cu(s)$

1. Place about 3 cm^3 of potassium manganate(VII) solution in a test-tube and add about 1 cm^3 of dilute sulphuric acid.

2. Mix the contents and add a <u>very</u> small amount of copper powder (the smallest amount on the tip of a spatula).

3. Cork and shake the tube well and note any change.

F. Reaction between $S_2O_8^{2-}(aq)$ and $Cr^{3+}(aq)$

1. Place about 3 cm^3 of potassium peroxodisulphate(VI) solution in a test-tube and add about 3 cm^3 of chromium(III) chloride solution.

2. Cork and shake the tube and note any change. Complete Results Table 47.

Question

Do your results agree with predictions made using $E^\ominus$ values? (See below.)

Half-cell reaction	$E^\ominus$/V
$Zn^{2+}(aq) + 2e^- \rightleftharpoons Zn(s)$	-0.76
$Cu^{2+}(aq) + 2e^- \rightleftharpoons Cu(s)$	+0.34
$I_2(aq) + 2e^- \rightleftharpoons 2I(aq)$	+0.54
$Fe^{3+}(aq) + e^- \rightleftharpoons Fe^{2+}(aq)$	+0.77
$Br_2(aq) + 2e^- \rightleftharpoons 2Br^-(aq)$	+1.07
$Cr_2O_7^{2-}(aq) + 14H^+(aq) + 6e^- \rightleftharpoons 2Cr^{3+}(aq) + 7H_2O(l)$	+1.33
$Cl_2(aq) + 2e^- \rightleftharpoons 2Cl^-(aq)$	+1.36
$MnO_4^-(aq) + 8H^+(aq) + 5e^- \rightleftharpoons Mn^{2+}(aq) + 4H_2O(l)$	+1.51
$S_2O_8^{2-}(aq) + 2e^- \rightleftharpoons 2SO_4^{2-}(aq)$	+2.01

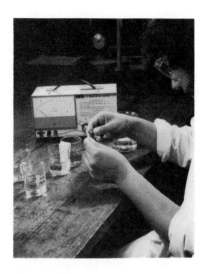

Aim

The purpose of this experiment is to investigate the effect of changes in silver ion concentration on the potential of the silver electrode.

Introduction

The cell you study in this experiment is represented as follows:

$$Cu(s) \mid Cu^{2+}(aq, 1.0 \text{ M}) \vdots Ag^{+}(aq, x \text{ M}) \mid Ag(s)$$

You vary the concentration of silver ions (x mol dm^{-3}) in the silver half-cell, whilst keeping the copper half-cell unchanged. The measured e.m.f. enables you to calculate the electrode potential of the $Ag^{+} \mid Ag$ electrode for each concentration of silver ions.

Requirements

safety spectacles
copper foil
silver wire
7 beakers, 50 cm^{3}
6 strips of filter paper
saturated potassium nitrate solution, 3 M KNO_3
voltmeter (high resistance)
2 connecting leads, with crocodile clips attached
copper(II) sulphate solution, 1.0 M $CuSO_4$
silver nitrate solutions of six different concentrations,
0.10 M, 0.010 M, 0.0033 M, 0.0010 M, 0.00033 M, 0.00010 M $AgNO_3$
silver residues bottle

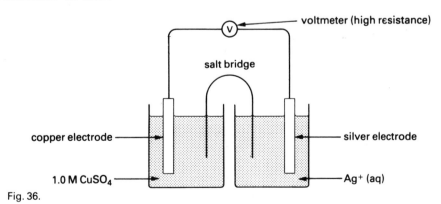

Fig. 36.

113

Procedure

1. Set up the cell in Fig. 36, using 1.0 M $CuSO_4$ and 0.00010 M $AgNO_3$.

2. Measure the potential difference of the cell and note the polarity of the electrodes - this will enable you to give a sign to the cell e.m.f., ΔE. Record the results in a copy of Results Table 48. Remove the salt bridge as soon as possible.

3. Replace the beaker of 0.00010 M $AgNO_3$ with a beaker of 0.00033 M $AgNO_3$, renew the salt bridge and measure the e.m.f. of the cell.

4. Repeat this procedure for the other silver solutions, following the order in the table (i.e. the most dilute first).

Calculations

1. Calculate $\log([Ag^+(aq)]/mol\ dm^{-3})$ and enter the values in your table.

2. Calculate and record the electrode potential of the $Ag^+(aq)|Ag(s)$ electrode for each concentration of silver ions.

3. Plot a graph of $E(Ag^+|Ag)/V$ (y-axis) against $\log[Ag^+(aq)]$ (x-axis). Draw the y-axis from -0.1 to 1.0 and the x-axis from 0 to -15 so that you can use the graph later for a greater range than is covered in this experiment.

Results Table 48

$[Ag^+(aq)]/mol\ dm^{-3}$	$\log([Ag^+(aq)]/mol\ dm^{-3})$	$\Delta E/V$	$E(Ag^+|Ag)/V$
0.00010			
0.00033			
0.0010			
0.0033			
0.010			
0.10			

Questions

1. How can Le Chatelier's principle be applied to the variation in electrode potential with concentration?

2. Determine the slope of the graph. What does the slope correspond to in the Nernst equation?

EXPERIMENT 49

Measuring the solubility products
of some silver salts

Aim

The purpose of this experiment is to determine,
from e.m.f. measurements, the solubility
products of the following sparingly soluble
salts: AgCl, AgBr, AgI, $AgIO_3$.

Introduction

For most purposes, salts such as silver chloride are said to be insoluble in
water. However, no substance is completely insoluble. Silver chloride is
in equilibrium with its ions in aqueous solution, and the solubility product,
K_s, is expressed by reference to the equation:

$$AgCl(s) \rightleftharpoons Ag^+(aq) + Cl^-(aq); \quad K_s = [Ag^+(aq)][Cl^-(aq)]$$

In this experiment you prepare mixtures containing solid silver salts in
equilibrium with very small concentrations of silver ions and a known excess
of the anions. You use each mixture to construct a cell and from the
measurement of its e.m.f. you determine the concentration of silver ions
using the graph plotted in Experiment 48.

Requirements

safety spectacles
5 beakers, 50 cm³
measuring cylinder, 25 cm³
strip of copper foil
piece of silver wire
filter paper strips
emery paper
voltmeter, high resistance
2 leads with crocodile clips
copper(II) sulphate solution, 1.0 M $CuSO_4$
silver nitrate solution, 0.10 M $AgNO_3$
potassium chloride solution, 0.10 M KCl
potassium bromide solution, 0.10 M KBr
potassium iodide solution, 0.10 M KI
potassium iodate(V) solution, 0.10 M KIO_3
potassium nitrate solution, saturated, 3 M KNO_3
silver residues bottle

Procedure

1. Prepare one of the systems 1 to 4 by swirling together in a smaller beaker the solutions shown in Table 49a. In view of the high cost of silver solutions, your teacher may limit each student to one system only.

Table 49a

	Volumes of solutions to make mixture	
System	0.10 M $AgNO_3$	Solution of anion
1. $AgCl(s)$ ⇌ $Ag^+(aq)$ + $Cl^-(aq)$	10 cm³	20 cm³ of 0.10 M KCl
2. $AgBr(s)$ ⇌ $Ag^+(aq)$ + $Br^-(aq)$	10 cm³	20 cm³ of 0.10 M KBr
3. $AgI(s)$ ⇌ $Ag^+(aq)$ + $I^-(aq)$	10 cm³	20 cm³ of 0.10 M KI
4. $AgIO_3(s)$ ⇌ $Ag^+(aq)$ + $IO_3^-(aq)$	10 cm³	20 cm³ of 0.10 M KIO_3

2. Set up the cell:

$$Cu(s) \mid Cu^{2+}(aq, 1.0\ M) \vdots Ag^+(aq,\ x\ M) \mid Ag(s)$$

using the mixture prepared in step 1 in the silver half-cell and 1.0 M $CuSO_4$ in the copper half-cell. Use a salt bridge made by dipping a strip of filter paper in potassium nitrate solution.

3. Connect the cell to a high resistance voltmeter and measure the potential difference between the electrodes. Note the polarity of the silver electrode and record the value of ΔE in a copy of Results Table 49b.

4. If your teacher agrees, repeat the procedure from stages 1 to 3 with each system in turn. Make sure that the silver electrode is cleaned and you use a fresh salt bridge each time. If you do not investigate all four systems yourself, obtain results from other students.

Results Table 49b

				$[Ag^+(aq)]$	$[X^-(aq)]$	K_S/mol² dm⁻⁶	
System	ΔE/V	E/V	$\log[Ag^+(aq)]$	/mol dm⁻³	/mol dm⁻³	Expt.	Data
1. AgCl							
2. AgBr							
3. AgI							
4. $AgIO_3$							

Calculations

1. Calculate the potential, E, of the Ag^+/Ag half-cell for each system.

2. Use the graph from Experiment 48 to determine the equilibrium silver ion concentration in systems 1 to 4 (to two significant figures).

3. Calculate the equilibrium concentration, $[X^-(aq)]$, for each anion.

4. Use these results to calculate the solubility product of each silver salt and compare your values with those quoted in your data book.

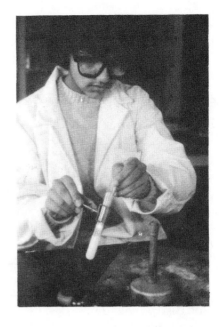

EXPERIMENT 50
Reaction between sodium peroxide and water

Aim

The purpose of this experiment is to identify
the products formed when sodium peroxide,
Na_2O_2, reacts with water.

Introduction

When the *s*-block monoxides dissolve in water, they produce hydroxide ions
which make the resulting solutions alkaline. For example,

$$CaO(s) + H_2O(l) \rightarrow Ca(OH)_2(aq)$$

In this experiment, you will note the pH of the resulting solution when sodium
peroxide, Na_2O_2, reacts with water, and identify other products which are also
formed.

One of the products you will be asked to test for is hydrogen peroxide, H_2O_2.
A very sensitive test for this is one which involves shaking it with orange
potassium dichromate(VI) solution, $K_2Cr_2O_7$, dilute sulphuric acid and
pentan-1-ol, $C_5H_{11}OH$. If a blue colour develops in the organic layer, then
hydrogen peroxide is present.

(The blue colour is thought to be due to CrO_5, which is stable in pentanol but
not in water.)

Requirements

safety spectacles
3 test-tubes in rack
Bunsen burner and bench mat
spatula
sodium peroxide, Na_2O_2 – – – – – – – – – – – – – – – – – –
distilled water
wood splints
3 dropping pipettes
universal indicator
potassium dichromate(VI) solution, 0.02 M $K_2Cr_2O_7$ – – – – – – – – –
dilute sulphuric acid, 1 M H_2SO_4
pentan-1-ol(amyl alcohol), $C_5H_{11}OH$ – – – – – – – – – – – –

Hazard warning

Pentanol is highly flammable. Therefore you MUST:

KEEP THE STOPPER ON THE BOTTLE WHEN NOT IN USE.
KEEP THE LIQUID AWAY FROM FLAMES.

Sodium peroxide is corrosive and a powerful oxidant.
Therefore you MUST:

WEAR SAFETY SPECTACLES   AVOID CONTACT WITH SKIN

Procedure

1. Cautiously add about 0.1 g of sodium peroxide to about 3 cm³ of distilled water in a test-tube, and immediately test the gas evolved. (What are the possibilities?)

2. To the resulting solution add a few drops of universal indicator.

3. Put a few grains (less than 0.1 g) of sodium peroxide in a test-tube and add the following reagents, in order: 3 cm³ of distilled water, 3 drops of potassium dichromate solution, 1 cm³ of pentan-1-ol and 3 cm³ of dilute sulphuric acid. Shake gently and let the two layers separate.

4. Record your observations and conclusions in a copy of Results Table 50.

Results Table 50

Experiment	Observation	Conclusion
Sodium peroxide added to water and universal indicator	Colour of gas .... Effect on glowing splint .... pH of solution	The gas is ....
Sodium peroxide plus water, added to acidified dichromate solution plus pentanol	Colour of organic (top) layer .... Colour of aqueous (bottom) layer	The colour of the pentanol layer ....

Questions

1. At 0°C, sodium peroxide reacts with water without the evolution of oxygen. Hydrogen peroxide is detected at this temperature.

   (a) Write an equation for this reaction.

   (b) The evolution of oxygen at higher temperatures is believed to be due to a secondary reaction. Write an equation for this secondary reaction.

   (c) What products are likely to be obtained if sodium peroxide is added to very hot water? Give a balanced equation.

2. (a) Predict the reaction between barium peroxide and ice-cold water. Give a balanced equation in your answer.

   (b) Hydrogen peroxide can be prepared in the laboratory by adding barium peroxide to ice-cold dilute sulphuric acid.

   (i) Write an equation for this reaction.

   (ii) Why do you think dilute sulphuric acid is used in the place of water?

3. All s-block oxides (apart from BeO) are basic oxides and thus react with water to form hydroxides, and with acidic substances to form salts. How would you expect the following oxides to react with carbon dioxide?

   (i) Na_2O,   (ii) Na_2O_2,   (iii) KO_2.

118

EXPERIMENT 51

Heating the nitrates and carbonates of the *s*-block elements.

Aim

The aim of this experiment is two-fold: to identify the products of thermal decomposition, and to estimate the order of thermal stability of these compounds.

Introduction

In this experiment you heat small samples of the nitrate and carbonate of each element, using the same size flame. You then note the time taken to detect the products of decomposition. Your pre-A-level experience should enable you to recognise nitrogen dioxide, NO_2 (brown gas), produced from nitrates, and carbon dioxide, CO_2, from carbonates. In order to save time, we suggest that you work in pairs, one student on each part of the experiment.

Requirements

safety spectacles
30 test-tubes
2 test-tube racks
labels for test-tubes
solid nitrates and carbonates (anhydrous) of Groups I and II
2 spatulas
retort stand, clamp and boss
bent delivery tube to fit test-tubes
lime-water
2 Bunsen burners and mats
2 stop-clocks (or watches)
2 test-tube holders
wood splints
hydrochloric acid, dilute, 2 M HCl

Hazard warning

Barium compounds are poisonous if swallowed or absorbed through the skin.

AVOID CONTACT WITH SKIN.

Nitrogen dioxide is a poisonous gas.

HEAT NITRATES IN A FUME CUPBOARD.

Nitrates are strong oxidizing agents.

DO NOT ALLOW PIECES OF GLOWING SPLINT TO DROP ON TO HOT NITRATES.

Procedure A. Effect of heat on carbonates

1. Set up two rows of four or five test-tubes each.

2. Label the test-tubes in the first row with the names of the Group I carbonates, and the second row with the names of the Group II carbonates.

3. To each test-tube add a spatula-measure of the appropriate carbonate.

4. For the first carbonate, set up the apparatus as shown in Fig. 37.

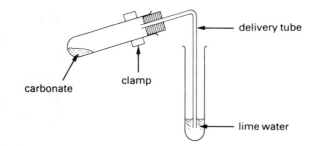

Fig. 37.

5. Start the clock at the moment you begin heating the carbonate by holding the end of the tube just above the inner blue cone of a roaring Bunsen flame.

6. Note the time for the lime-water to just turn milky (if it does at all) and, before removing the flame, prevent 'suckback' by lifting the delivery tube out of the lime-water.

7. Repeat the heating procedure for each carbonate in turn, making sure that the tube height and the flame size are the same for each one.

8. Record your results in a copy of Results Table 51a.

Results Table 51a. Effect of heat on the s-block carbonates

Carbonate	Time to detect CO_2	Observations
Li_2CO_3		
Na_2CO_3		
K_2CO_3		
Rb_2CO_3*		
Cs_2CO_3*		
$MgCO_3$		
$CaCO_3$		
$SrCO_3$		
$BaCO_3$		

*If available

Procedure B. Effect of heat on nitrates. (To be done in a fume cupboard)

9. Heat small quantities of sodium nitrate and magnesium nitrate separately, in clean test-tubes. In each case, test for oxygen and note the appearance of any brown fumes. (This will give you an idea of the products you can expect when Groups I and II nitrates are heated.)

10. Set up two rows of four test-tubes each containing the appropriate nitrates, as you did for the carbonates.

11. Start the clock at the moment you begin heating the first nitrate by holding the end of the test-tube just above the blue cone of a roaring Bunsen flame.

12. Test for oxygen at short regular intervals (place the glowing splint in the same part of the test-tube) and also look for the first signs of a brown gas. (A white background is helpful.)

13. In a copy of Results Table 51b, record the time for the first appearance of brown fumes or when oxygen is detected. (Whichever method of detection you choose for one member of a group you must also apply to the other members of the same group.) Continue heating for another minute.

14. Repeat steps 11, 12 and 13 for the other nitrates in turn.

15. To the cold solid residue remaining after each nitrate is heated add a few drops of dilute hydrochloric acid and warm.

Results Table 51b   Effect of heat on the s-block nitrates

Nitrate	Time to detect O_2 or NO_2	Observations	Effect of adding dilute HCl to cold residue
$LiNO_3$			
$NaNO_3$			
KNO_3			
$RbNO_3$*			
$CsNO_3$*			
$Mg(NO_3)_2$			
$Ca(NO_3)_2$			
$Sr(NO_3)_2$			
$Ba(NO_3)_2$			

*If available

Questions

1. Explain why many of these nitrates rapidly turn into colourless liquids on first heating, whilst on further heating become white solids again, before they decompose. (Hint: look at the formulae of the nitrates on the reagent bottles.)

2. Which of the Group I nitrates and carbonates most resemble those of Group II in their reaction to heat?

3. Describe the trend in thermal stability of the nitrates and carbonates as each group is descended.

4. How do Group I nitrates and carbonates compare with their Group II counterparts in terms of thermal stability?

5. Write balanced equations, where applicable, for the thermal decompositions of:

   (a) lithium nitrate and lithium carbonate;

   (b) potassium nitrate and potassium carbonate;

   (c) magnesium nitrate and magnesium carbonate.

6. Why are brown fumes produced when dilute HCl is added to the cold residue obtained when many of Group I nitrates are heated?

EXPERIMENT 52

The solubility of some salts of
Group II elements

Aim

The aim of this experiment is to demonstrate
the trends in solubility of the Group II
carbonates, sulphates, sulphites and hydroxides.

Introduction

In this experiment, you add each of the anion solutions to 1 cm³ of each
cation solution provided, drop by drop, until the first sign of a precipitate
appears. For each salt, the solubility is proportional to the number of drops
of anion added.

Requirements

safety spectacles
16 test-tubes and four racks
labels for test-tubes
5 teat-pipettes (marked off at 1 cm³)
0.1 M solutions of the following cations: Mg^{2+}, Ca^{2+}, Sr^{2+}, Ba^{2+}
1.0 M solution of OH^-
0.5 M solutions of $SO_4{}^{2-}$ and $SO_3{}^{2-}$ ions
0.05 M solution of $CO_3{}^{2-}$ ions
distilled water

Hazard warning

Barium compounds are poisonous if swallowed
or absorbed through the skin.

Procedure

1. Set up four rows of four test-tubes each.

2. For each row, label the first test-tube Mg^{2+}, the second test-tube Ca^{2+},
    the third test-tube Sr^{2+} and the fourth test-tube Ba^{2+}.

3. Add 1 cm³ of the appropriate cation solution to each test-tube, using a
    teat-pipette with a 1 cm³ mark.

4. Label the first row of test-tubes OH^-, the second row $SO_4{}^{2-}$, the third
    row $SO_3{}^{2-}$, and the fourth row $CO_3{}^{2-}$.

5. Add the OH^-, drop by drop, with shaking, to each cation solution in the
    first row, until the first sign of a precipitate appears.

6. Record the number of drops of OH^- solution used in a copy of Results
    Table 52.

7. Repeat Steps 5 and 6 with the remaining anions and cations.

8. If a precipitate appears suddenly, during the addition of a drop, then
    you should classify the precipitate as slight (s) or heavy (h).

9. If no precipitate appears after forty drops, then write '40+' and regard
    the salt as soluble.

Results Table 52

Cation solution	Number of drops of anion solution added to give a precipitate			
	OH^-	SO_4^{2-}	SO_3^{2-}	CO_3^{2-}
Mg^{2+}				
Ca^{2+}				
Sr^{2+}				
Ba^{2+}				

Question

For Group II, what are the trends in solubility of the salts listed below?

(a) hydroxides;   (c) sulphites;

(b) sulphates;   (d) carbonates.

EXPERIMENT 53

The solubility of the halogens
in organic solvents

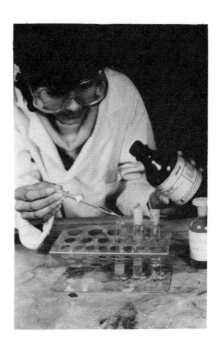

Aim

The purpose of this experiment is to discover
whether each of the halogens chlorine, bromine
and iodine is:

(a)   soluble in organic solvents,

(b)   more soluble in organic solvents than in
      water,

(c)   the same colour in organic solvents as
      it is in water.

Introduction

After mixing aqueous solutions of the halogens separately with 1,1,1-tri-
chloroethane, CH_3CCl_3, ethoxyethane, $CH_3CH_2OCH_2CH_3$ (ether) and hexane,
$CH_3CH_2CH_2CH_2CH_2CH_3$, you decide whether each halogen moves out of the aqueous
solution and into the solvent. If it does, then you can compare its colour
in the aqueous layer with that in the organic layer - these colours are
useful for identifying the halogens.

Requirements

safety spectacles
3 test-tubes, each fitted with a bung
test-tube rack
6 dropping pipettes approximately graduated for 1 cm³
1,1,1-trichloroethane, CH_3CCl_3 — — — — — — — — — — — — — — — —
ethoxyethane (diethyl ether), $CH_3CH_2OCH_2CH_3$ — — — — — — — — —
hexane, $CH_3CH_2CH_2CH_2CH_2CH_3$ — — — — — — — — — — — — — — — — —
chlorine water, $Cl_2(aq)$ — — — — — — — — — — — — — — — — —
bromine water, $Br_2(aq)$ — — — — — — — — — — — — — — — — — —
iodine solution, 0.01 M I_2 (in KI(aq))
3 labelled bottles for organic residues

Hazard warning

This experiment should be performed in a FUME CUPBOARD since all
the halogen vapours are TOXIC and the organic vapours are
dangerous to inhale. If a fume cupboard is not available then
THE LABORATORY MUST BE WELL VENTILATED.

Check that no flames are close to where you are working because
ethoxyethane and hexane are extremely flammable.

Procedure

1. In a copy of Results Table 53 note the colour of each aqueous halogen solution provided.

2. Into each of three test-tubes in turn put about 1 cm³ of a different aqueous halogen solution.

3. To each tube add about 1 cm³ of 1,1,1-trichloroethane and note whether the organic liquid becomes the upper or lower layer.

4. Cork and shake each tube, allow the layers to separate, and note the colour of each layer.

5. Repeat steps 2, 3 and 4 with the other two solvents.

6. Pour residues into the labelled bottles provided, not down the sink.

Results Table 53

	Nature of each layer	Chlorine water	Bromine water	Iodine solution
Colour of aqueous solution.				
Colour of each layer after shaking with 1,1,1-trichloroethane.	Upper layer *organic/aqueous			
	Lower layer *organic/aqueous			
Colour of each layer after shaking with ethoxyethane.	Upper layer *organic/aqueous			
	Lower layer *organic/aqueous			
Colour of each layer after shaking with hexane.	Upper layer *organic/aqueous			
	Lower layer *organic/aqueous			

*delete organic or aqueous as appropriate

Questions

1. Do you think the halogens are more soluble in the organic solvents than in water? Explain your answer.

2. Which of the three halogens has a significantly different colour in the organic layer from its colour in the aqueous layer? (Specify the organic solvents in your answer.)

3. How would you distinguish, other than by smell, between a dilute aqueous iodine solution and a fairly concentrated solution of bromine water?

EXPERIMENT 54

The action of dilute alkali
on the halogens

Aim

The purpose of this experiment is to show
that observable reactions occur between
dilute sodium hydroxide and aqueous
solutions of bromine and iodine and,
furthermore, that these reactions are
reversible.

Introduction

You add dilute sodium hydroxide dropwise to bromine water and iodine solution
in turn. A significant change in colour in either halogen solution indicates
that a reaction has occurred. If you suspect that the change in colour is
only due to a dilution effect, you should set up a control experiment, where
you add the same number of drops of distilled water to the halogen solution.

To determine whether the reaction is reversible, you acidify the alkaline
halogen solution and see if the original halogen colour reappears.

Requirements

safety spectacles
5 test-tubes
1 test-tube rack
4 teat-pipettes
bromine water, $Br_2(aq)$ —
iodine solution, 0.01 M I_2 (in KI(aq))
sulphuric acid, 1 M H_2SO_4
sodium hydroxide solution, 2 M NaOH — — — — — — — — — — — — — — — — —
distilled water

Hazard warning

Bromine water is poisonous and corrosive. The vapour is extremely
irritant to the eyes, lungs and skin. Therefore you must:
AVOID CONTACT WITH SKIN;
AVOID INHALING THE VAPOUR.

Procedure

1. Place about 2 cm³ of bromine water in a test-tube and note the colour of
   the solution.

2. Add dilute sodium hydroxide, drop by drop, and note any change of colour
   in the bromine solution.

3. Now add dilute sulphuric acid to the solution from (2), and note if the
   colour returns when the acid is in excess.

4. Repeat the procedure using iodine solution instead of bromine water.

5. Record your results in a copy of Results Table 54.

Results Table 54

Aqueous halogen	Original colour	Colour after adding NaOH(aq)	Colour after adding H_2SO_4(aq)
Bromine water			
Iodine solution			

Questions

1. Write equations for the reactions of bromine and iodine with dilute sodium hydroxide at room temperature. Use your text-book(s) as necessary.

2. Which of the above reactions would you class as disproportionation reactions?

3. Are these reactions reversible? Explain your answer.

4. How does chlorine differ from bromine and iodine in its reaction with cold dilute sodium hydroxide? Suggest a reason.

EXPERIMENT 55
Halogen-halide reactions
in aqueous solution

Aim

The purpose of this experiment is to investigate the order of oxidizing ability of the halogens Cl_2, Br_2 and I_2 in aqueous solution.

Introduction

You mix each of the aqueous solutions with halide ion solutions, $Cl^-(aq)$, $Br^-(aq)$, and $I^-(aq)$ in turn, and see whether a reaction takes place. The addition of hexane to the halogen-halide mixture enables you to recognise the halogen molecules present. The halogen which oxidizes most of the other halide ions will clearly be the strongest oxidizing agent.

Requirements

safety spectacles
6 test-tubes fitted with corks
test-tube rack
7 dropping pipettes
bromine water, $Br_2(aq)$ —
chlorine water, $Cl_2(aq)$ —
iodine solution, 0.01 M I_2 (in KI(aq))
potassium bromide solution, 0.1 M KBr
potassium chloride solution, 0.1 M KCl
potassium iodide solution, 0.1 M KI
hexane, C_6H_{14} —
bottle for residues

Hazard warning

Halogen and organic vapours must not be inhaled. If a fume cupboard is not available then THE LABORATORY MUST BE WELL VENTILATED AND REAGENT BOTTLES AND TEST-TUBES STOPPERED AS MUCH AS POSSIBLE. KEEP HEXANE AWAY FROM FLAMES.

Procedure

1. Reaction (if any) of iodide with chlorine and bromine.

    (a) To each of two test-tubes add about 1 cm³ of potassium iodide solution.

    (b) To one of these tubes, add about the same volume of chlorine water, and to the other add the same volume of bromine water.

    (c) Cork and shake the tubes and note the colour change - if any.

    (d) To each tube add about 1 cm³ of hexane, cork and shake, allow to settle, and note the colour of each layer.

    (e) Decide which reactions have taken place, and complete a copy of Results Table 55.

2. Reaction (if any) of bromide with chlorine and iodine

Repeat the above steps, 1(a) - (e), using potassium bromide with chlorine water and iodine solution.

3. Reaction (if any) of chloride with bromine and iodine

Repeat steps 1(a) - (e) using potassium chloride with bromine water and iodine solution.

Results Table 55

			Chlorine water	Bromine water	Iodine solution
	Initial colour				
1.	Colour after shaking with KI solution				
	Colour of each layer after shaking with hexane	Upper			
		Lower			
	Conclusion				
2.	Colour after shaking with KBr solution				
	Colour of each layer after shaking with hexane	Upper			
		Lower			
	Conclusion				
3.	Colour after shaking with KCl solution				
	Colour of each layer after shaking with hexane	Upper			
		Lower			
	Conclusion				

Questions

1. (a) Does I_2(aq KI) oxidize Cl^-(aq) and Br^-(aq)?

   (b) Does Br_2(aq) oxidize Cl^-(aq) and I^-(aq)?

   (c) Does Cl_2(aq) oxidize Br^-(aq) and I^-(aq)?

2. Write ionic equations for the reactions taking place.

EXPERIMENT 56
Reactions of solid halides

Aim

The purpose of this experiment is to study
the effect of an oxidizing acid (concentrated
sulphuric acid) and a non-oxidizing acid
(phosphoric(V) acid) on three solid potassium
halides; potassium chloride, potassium bromide
and potassium iodide.

Introduction

In this experiment, you mix separate samples of crystalline potassium chloride,
potassium bromide and potassium iodide with the following reagents in turn:

(a) $H_2SO_4(l)$ and $MnO_2(s)$ (b) $H_2SO_4(l)$ alone   (c) $H_3PO_4(l)$

Possible products include the halogens, the hydrogen halides, sulphur
dioxide (SO_2) and hydrogen sulphide (H_2S). You already know from
Experiment 55 how to recognise the halogens. Tests to recognise the
other products are as follows:

Hydrogen halides. Hold a moist stopper from a bottle of ammonia solution
near the source of the gas. Dense white fumes indicate the presence of a
hydrogen halide (or other strongly-acidic gas).

Sulphur dioxide. Hold a strip of filter paper soaked in acidified potassium
dichromate(VI) solution near the source of the gas. A colour change from
orange to green indicates the presence of sulphur dioxide (or other strongly-
reducing gas).

Hydrogen sulphide. The 'bad egg' smell is very characteristic, but take care
- the gas is very toxic. Hold a strip of filter paper soaked in lead
ethanoate (acetate) solution near the source of the gas. A silver black
colour indicates the presence of hydrogen sulphide.

Requirements

safety spectacles
access to fume cupboard
12 test-tubes in rack
spatula
potassium bromide, solid, KBr
potassium chloride, solid, KCl
potassium iodide, solid, KI
manganese(IV) oxide, MnO_2
1 glass stirring rod
sulphuric acid, concentrated, H_2SO_4 $------------------$
phosphoric(V) acid, 100%, H_3PO_4 $---------------$
test-tube holder
Bunsen burner and bench mat
distilled water
3 dropping pipettes
ammonia solution, 0.880 NH_3 $------------------$
lead(II) ethanoate solution, $(CH_3COO)_2Pb$ $----------------$
potassium dichromate(VI) solution, $K_2Cr_2O_7$ $----------------$
strips of filter paper
starch solution
hexane, C_6H_{14} $-----------------------$

Hazard warning

Concentrated sulphuric acid is very corrosive and reacts violently
with water. Phosphoric(V) acid and potassium dichromate are also
very corrosive. Therefore, you MUST:

AVOID CONTACT WITH SKIN; if contact does occur, wash immediately under a
cold tap with <u>plenty</u> of water.

DISPOSE OF <u>COLD</u> RESIDUES CONTAINING CONCENTRATED SULPHURIC ACID BY POURING
<u>SLOWLY</u> INTO PLENTY OF WATER.

The halogens, hydrogen halides and hydrogen sulphide are toxic.
Therefore you must CARRY OUT THESE EXPERIMENTS IN A FUME CUPBOARD.

Hexane is very flammable. KEEP HEXANE STOPPERED AND AWAY FROM FLAMES.

Procedure

1. Reaction with H_2SO_4 and MnO_2

    (a) Into three separate test-tubes, place enough potassium chloride,
         potassium bromide and potassium iodide to half-fill the rounded
         part at the bottom.

    (b) To the contents of each test-tube, add a roughly equal quantity of
         manganese(IV) oxide, and mix the solids together with a stirring
         rod.

    (c) Hold the tube in a fume cupboard, with its mouth pointed away from
         you, and cautiously add ten drops of concentrated sulphuric acid,
         shaking the tube gently after the addition of each drop.

    (d) Note whether any reaction occurs, and confirm any suspected products
         by appropriate tests. Complete a larger copy of Results Table 56.

    (e) If no reaction seems to occur, warm the test-tube carefully.

2. Reaction with H_2SO_4

    Repeat the above procedure without using manganese(IV) oxide.

3. Reaction with H_3PO_4

Repeat the above procedure, using phosphoric(V) acid alone in place of sulphuric acid, i.e. without using manganese(IV) oxide.

Results Table 56

Test	Chloride	Bromide	Iodide
1. Action of conc. H_2SO_4 and MnO_2 Observations Suspected product(s) Confirmatory tests			
2. Action of conc. H_2SO_4 Observations Suspected product(s) Confirmatory tests			
3. Action of H_3PO_4 Observations Suspected product(s) Confirmatory tests			

Questions

1. In many of the reactions you may have detected mixtures of the halogens and the hydrogen halides. In such cases, you should assume that at least two reactions are occurring. With this in mind and with the aid of text-books, complete and balance the following equations:

(a) $KCl(s) + H_2SO_4(l) \rightarrow$

(b) $KCl(s) + H_3PO_4(l) \rightarrow$

(c) $KBr(s) + H_2SO_4(l) \rightarrow$

   $HBr(g) + H_2SO_4(l) \rightarrow$

(d) $KBr(s) + H_3PO_4(l) \rightarrow$

(e) $KI(s) + H_2SO_4(l) \rightarrow$

   $HI(g) + H_2SO_4(l) \rightarrow$

(f) $KI(s) + H_3PO_4(l) \rightarrow$

(We have used the formula $H_3PO_4(l)$ rather than $H_3PO_4(s)$ because the solid melts before reaction occurs.)

2. Manganese(IV) oxide is a strong oxidizing agent capable of oxidizing all the hydrogen halides (except HF) to the halogens. In the light of this statement, explain the reactions between potassium chloride and concentrated sulphuric acid, with and without manganese(IV) oxide.

3. Why does the addition of manganese(IV) oxide appear to have little effect on the reaction between potassium iodide and concentrated sulphuric acid?

133

EXPERIMENT 57
Reaction of halides in solution

Aim

The purpose of this experiment is to find out whether the ions, Cl⁻, Br⁻ and I⁻, react in solution with certain reagents and, where they do react, what products are formed.

Introduction

In this experiment, you add various reagents to separate samples of solutions containing the Cl⁻, Br⁻ and I⁻ ions. In many of the reactions, precipitates are formed. Where you are asked to add another reagent to excess, you should look carefully to see if any of the precipitate dissolves.

Requirements

safety spectacles
18 test-tubes, 9 with corks
3 test-tube racks
8 dropping pipettes
potassium bromide solution, 0.1 M KBr
potassium chloride solution, 0.1 M KCl
potassium iodide solution, 0.1 M KI
silver nitrate solution, 0.02 M AgNO₃
nitric acid, dilute, 2 M HNO₃
ammonia solution, 15 M NH₃
lead(II) nitrate solution, 0.1 M Pb(NO₃)₂
hydrogen peroxide solution, H₂O₂
starch solution, 1%
sulphuric acid, dilute, 1 M H₂SO₄
hexane, C₆H₁₄

Hazard warning

KEEP HEXANE WELL STOPPERED AND AWAY FROM FLAMES.

Procedure

Add the following reagents to 1 cm³ of the chloride, bromide and iodide solutions in turn, and record your observations in a copy of Results Table 57.

1. Add approximately 1 cm³ of silver nitrate solution and shake gently. Note what happens. Move the three tubes to a dark cupboard, leave them there till the end of the lesson and note their appearance again.

2. Add silver nitrate solution as in (1). Leave these tubes in their racks until the end of the lesson, noting their appearance every 10-15 minutes.

3. Add approximately 1 cm³ of silver nitrate solution followed by excess (e.g. 5 cm³) dilute nitric acid. Cork the test-tube and shake vigorously.

4. Add approximately 1 cm³ silver nitrate solution followed by excess (e.g. 5 cm³) ammonia solution. Cork the test-tube and shake.

5. Add approximately 1 cm³ lead(II) nitrate solution.

6. Add approximately 1 cm³ hydrogen peroxide solution followed by approximately 1 cm³ dilute sulphuric acid. Cork these tubes and allow them to stand. Add any further reagent(s) which you think will help you to decide what has happened.

Results Table 57

Test	Chloride	Bromide	Iodide
Action of $AgNO_3$(aq) Effect of standing in (a) dark (b) light			
Action of $AgNO_3$(aq) followed by dilute HNO_3(aq)			
Action of $AgNO_3$(aq) followed by NH_3(aq)			
Action of $Pb(NO_3)_2$ (aq)			
Action of H_2O_2(aq) and dilute H_2SO_4 (aq)			

Questions

1. Write ionic equations for the reactions between each of the three halide solutions and

   (a) silver nitrate solution,

   (b) lead(II) nitrate solution.

2. What chemical tests would you perform in order to distinguish between

   (a) Cl^-(aq) and Br^-(aq),

   (b) Br^-(aq) and I^-(aq)?

3. (a) Write an ionic equation for the reaction between an aqueous iodide and acidified hydrogen peroxide.

   (b) Why do you think no reaction occurs between acidified hydrogen peroxide and the other halide ions?

4. Suggest a reason for the darkening effect of light on the silver chloride and silver bromide precipitates.

EXPERIMENT 58
Reactions of the halates

Aim

The purpose of this experiment is to show the effect of heat on salts of the halogen oxoacids and to compare their oxidizing properties.

Introduction

In this experiment you carry out some simple test-tube experiments on sodium chlorate(I), sodium chlorate(V) and sodium iodate(V). Sodium chlorate(I) exists only in solution whereas the others can be found as crystalline solids.

From the properties of these three salts you can make predictions about the salts of the other oxoacids.

Requirements

safety spectacles
7 test-tubes in rack
spatula
sodium chlorate(V), solid, $NaClO_3$ — — — — — — — — — — — — — — — — —
sodium iodate(V), solid, $NaIO_3$
test-tube holder
Bunsen burner and bench mat
wood splints
sodium chlorate(I) solution, 1 M NaClO
distilled water
universal indicator solution
potassium iodide solution, 0.1 M KI
sulphuric acid, dilute, 1 M H_2SO_4
cobalt chloride solution, 0.01 M $CoCl_2$
blue and red litmus papers

Procedure

1. Place about 0.1 g of sodium chlorate(V) in one test-tube and an equal quantity of sodium iodate(V) in another test-tube.

2. Heat each test-tube strongly and test the gas given off with a glowing splint.

    Record your observations in a larger copy of Results Table 58.

3. To 1 cm³ of an aqueous solution of each of the oxoacid salts add a few drops of the following reagents, and record your observations in a copy of Results Table 58.

    (a) Universal indicator solution.

    (b) Potassium iodide solution.

    (c) Potassium iodide solution acidified with dilute sulphuric acid.

    (d) Cobalt chloride solution. Warm and test any gas given off with a glowing splint.

    (e) Dilute sulphuric acid. Test any gas with damp blue litmus paper.

Results Table 58

	NaClO	NaClO$_3$	NaIO$_3$
Effect of heat on solid	░░░░░		
pH of solution			
Addition of neutral KI(aq)			
Addition of acidified KI(aq)			
Addition of Co^{2+}(aq)			
Addition of dilute H$_2$SO$_4$(aq)			

Questions

1. Use your text-book(s) to write balanced equations for the effect of heat on the two solid salts.

2. Name the common product formed when the salts react with acidified potassium iodide solution.

3. Which of the salts is the strongest oxidizing agent? Explain your answer.

4. How would you expect potassium bromate(V) to react with acidified potassium iodide solution?

5. Would you expect sodium chlorate(VII), NaClO$_4$, to be a powerful oxidizing agent?

6. In reaction 3(d), Co^{2+} ions act as a catalyst. Write an ionic equation for the decomposition of aqueous sodium chlorate(I).

7. Write an equation showing the effect of a dilute acid on sodium chlorate(I). (Hint: Cl$^-$ ions are always present in solutions of sodium chlorate(I) and they appear in the equation.)

EXPERIMENT 59
Balancing a redox reaction

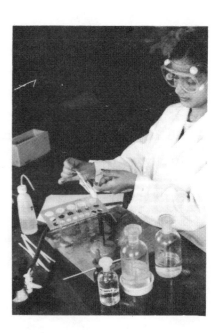

Aim

The purpose of this experiment is to calculate the amount of iodide ions which react with each mole of iodate(V) ions in aqueous solution.

Requirements

safety spectacles
2 measuring cylinders, 10 cm³
potassium iodide solution, ~ 1 M KI
2 conical flasks, 250 cm³
hydrochloric acid, ~ 2 M HCl
2 burettes, stands and filter funnels
1 white tile
potassium iodate solution, 0.10 M KIO_3
sodium thiosulphate solution, 0.10 M $Na_2S_2O_3$
starch solution, 0.2%
wash-bottle of distilled water

Procedure

1. Use a measuring cylinder to pour about 10 cm³ of potassium iodide solution into a 250 cm³ conical flask.

2. To the solution in the conical flask add about 10 cm³ of dilute hydrochloric acid.

3. From a burette, add precisely 5.0 cm³ of 0.10 M potassium iodate solution to the acidified iodide solution.

4. Titrate the iodine formed against 0.10 M sodium thiosulphate solution. When the colour of the iodine has nearly gone, add 1-2 cm³ of starch solution and continue the addition of thiosulphate solution drop by drop until the blue colour disappears.

5. Record your burette readings in a copy of Results Table 59.

6. Repeat steps 1 to 5 as a check on your accuracy.

Results Table 59

Solution in flask					
Solution in burette				mol dm^{-3}	
Indicator					

		Trial	1	2	3	4
Burette readings	Final					
	Initial					
Volume used/cm^3						
Mean titre/cm^3						

Calculations

1. Calculate the amount of sodium thiosulphate present in the volume of solution run out from the burette.

2. Calculate the amount of iodine <u>atoms</u> which must have reacted with the amount of $S_2O_3{}^{2-}$(aq) calculated in step 1. Use the equation:

$$2S_2O_3{}^{2-}(aq) + I_2(aq) \rightarrow 2I^-(aq) + S_4O_6{}^{2-}(aq)$$

3. Calculate the amount of iodine <u>atoms</u> present in 5.0 cm^3 of 0.10 M KIO_3.

4. Subtract the value obtained in step 3 from the value obtained in step 2 to obtain the amount of iodine atoms which originated from the potassium iodide.

5. State the amount of iodide ions which reacts with each mole of iodate ions.

EXPERIMENT 60
Observation and deduction exercise

Aim

The purpose of this experiment is to give you some practice in the investigation of unknown substances.

Introduction

The procedure below is taken from an A-level practical examination paper; read it carefully and report fully.

A part

Requirements

safety spectacles
5 test-tubes in rack
test-tube holder
alkali metal salt, F
spatula
Bunsen burner and bench mat
red and blue litmus papers
wood splints
wash-bottle of distilled water
silver nitrate solution, 0.02 M $AgNO_3$ – – – – – – – – – – – – – – – –
nitric acid, dilute, 2 M HNO_3
chlorine water, Cl_2 –
lead(II) ethanoate solution, 0.1 M $(CH_3CO_2)_2Pb$ – – – – – – – – – – –
other chemicals, for testing gases, are available from your teacher

Procedure

You are provided with an alkali metal salt, F. Carry out the following tests and record your observations and inferences in (larger copies of) the tables provided. Then answer the question which follows the tables.

Results Table 60a

Test	Observations	Inferences
(a) Heat approximately 0.1 g of F in a pyrex tube, at first gently and then more strongly, until the change is complete. Cool and keep the residue. Test any gases evolved.		
(b) Make an aqueous solution of the residue from (a) and carry out the following tests on portions: (i) Add aqueous silver nitrate followed by dilute nitric acid. (ii) Add aqueous chlorine. (iii) Add aqueous lead(II) ethanoate (lead acetate)		

Now answer the following question:

For a non-metal in F, give two substances or ions involved in the reactions in (a) and (b) which contain the non-metal and in which the non-metal has different oxidation numbers. Write your answer in the following table.

Results Table 60b

Substance provided	Name of non-metal	Name and formula of substance ion	Oxidation number
F			

EXPERIMENT 61
Observation and deduction exercise

Aim

The purpose of this experiment is to give
you some practice in the investigation of
unknown substances.

Introduction

The procedure below is taken from an A-level practical examination
paper; read it carefully and report fully.

A

Requirements

safety spectacles
5 test-tubes in rack
test-tube holder
potassium salts, D and E
spatula
Bunsen burner and bench mat
filter-paper strips
red and blue litmus papers
wood splints
sulphuric acid, concentrated, H_2SO_4 — — — — — — — — — — — — — — — —
potassium manganate(VII) (permanganate) solution, 0.01 M $KMnO_4$
distilled water
sulphuric acid, dilute, 1 M H_2SO_4
other chemicals, for testing gases, are available from your teacher

---

Hazard warning

Concentrated sulphuric acid is very corrosive and reacts violently
with water. Therefore, you MUST:

AVOID CONTACT WITH SKIN; if contact does occur, wash immediately
under a cold tap with <u>plenty</u> of water.

DISPOSE OF <u>COLD</u> RESIDUES BY POURING <u>SLOWLY</u> INTO PLENTY OF WATER.

Procedure

You are provided with potassium salts, D and E. Test each salt in turn,
as follows.

(a) Heat a portion until any reaction ceases. Test any gases evolved.

(b) Allow the residue from (a) to cool; then cautiously add a few
     drops of concentrated sulphuric acid.

(c) Add a fresh portion of each salt to a few drops of aqueous potassium
     permanganate previously acidified with twice its volume of dilute
     sulphuric acid, and warm.

(d) Now make aqueous solutions of D and E and mix the two solutions.

(e) Acidify the mixture from (d) with dilute sulphuric acid.

Carefully observe what happens and report fully.

What tentative inferences do you draw from these experiments?

Carry out and report on TWO further experiments which test your inferences.
These experiments can be on D, E or on the products of the above reactions.

Full credit will not be given unless your answer discloses the method
(including the scale of your experiments), careful observations, and some
comment on the types of chemical reactions involved.

The record of your work must be made in the form of three tables.

Results Table 61a Tests with unknown substance D

Test	Method	Observations	Inferences
(a) Heat.			
(b) Concentrated sulphuric acid on cold residue from (a).			
(c) Acidified potassium manganate(VII) (permanganate) and warm.			

Results Table 61b Tests with unknown substance E

Test	Method	Observations	Inferences
(a) Heat.			
(b) Concentrated sulphuric acid on cold residue from (a).			
(c) Acidified potassium manganate(VII) (permanganate) and warm.			
(d) Mix aqueous solutions of D and E.			
(e) Dilute sulphuric acid with mixture from (d).			

Results Table 61c Experiments to test inferences

Inference tested	Test and observations	Conclusion

EXPERIMENT 62
Investigating the properties
of Period 3 chlorides

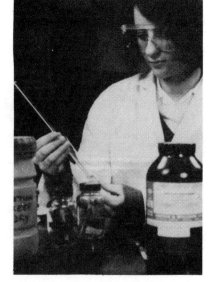

Aim

The purpose of this experiment is to study
the chlorides of Period 3 elements and
classify them according to structural type
and bonding.

Introduction

You first examine the appearance of each compound and then you find out
whether it dissolves in water and/or hexane. If it does dissolve you may
detect a temperature change. In general, a small temperature change
indicates a physical process and a large one a chemical process. This will
help you to distinguish between the physical process of dissolving and the
chemical one of hydrolysis when you add these substances to water.

You also determine any pH changes that take place when you mix the chlorides
with water. A decrease in pH indicates that hydrolysis has taken place.

Finally, you consider physical data for each compound and reach a conclusion
about its structure and bonding.

Requirements

safety spectacles
protective gloves
access to fume-cupboard
14 test-tubes (6 must be dry)
test-tube rack
2 measuring cylinders, 10 cm³ (1 must be dry)
distilled water
thermometer, 0-100 °C
spatula
universal indicator solution and colour chart
pH paper to cover the range 1 to 7
ammonia solution, 0.880 NH₃– – – – – – – – – – – – – – –
sodium chloride, NaCl
magnesium chloride, MgCl₂
aluminium chloride, AlCl₃ (anhydrous if possible) – – – – – –
4 teat-pipettes
silicon tetrachloride, SiCl₄– – – – – – – – – – – – – – – – – –
phosphorus trichloride, PCl₃– – – – – – – – – – – – – – –
disulphur dichloride, S₂Cl₂– – – – – – – – – – – – – – – – –
hexane, C₆H₁₄– –
organic residues bottle

144

Procedure

A. Appearance

Examine the chloride samples provided and, in a larger copy of Results
Table 62a, note for each:

(a) whether it is solid, liquid or gaseous,

(b) its colour (if any).

B. On mixing with water

1. Set up seven test-tubes, side by side.

2. Into each test-tube pour about 5 cm³ of distilled water.

3. In the first test-tube place a thermometer.

(a) Note the temperature.

(b) Add half a spatula-tip of sodium chloride and very carefully
stir with the thermometer.

(c) Note, after about one minute, (i) the temperature, (ii) whether
the solid has dissolved and (iii) anything else you see. For
example, is gas evolved at any time? If so, if the gas acidic?
Can you identify it using a simple test?

(d) Add 2-4 drops of universal indicator solution, or use a piece of
pH paper, compare the colour with the chart provided, and note
the pH indicated.

4. Repeat (but with more care!) the above steps 3.(a) - (d)
using, in turn, magnesium chloride, aluminium chloride, silicon
tetrachloride (2 drops), phosphorus trichloride (2 drops), and
disulphur dichloride (2 drops).

5. Measure the pH of the water in the seventh test-tube by
adding 2-4 drops of universal indicator solution or by
using pH paper, for comparison with the above.

C. On mixing with hexane

1. Set up another six test-tubes, side by side. These must be dry.

2. Into each test-tube pour about 5 cm³ of hexane.

3. In the first test-tube place a thermometer.

(a) Note the temperature.

(b) Add half a spatula-tip of sodium chloride and stir very carefully with the thermometer,

(c) Note, after about one minute, (i) the temperature, (ii) whether the solid dissolves and (iii) anything else you see.

(Dispose of hexane by pouring into the residue bottle provided.)

4. Repeat the above steps 3.(a) - (c) using, in turn, magnesium chloride, aluminium chloride, silicon tetrachloride (2 drops), phosphorus trichloride (2 drops), and disulphur dichloride (2 drops).

Results Table 62a

	NaCl	MgCl$_2$	AlCl$_3$	SiCl$_4$	PCl$_3$	S$_2$Cl$_2$
Appearance						
On mixing with water Initial temperature Final temperature Does it dissolve? pH of solution Other observation(s) (if any)						
On mixing with hexane Initial temperature Final temperature Does it dissolve? Other observation(s) (if any)						

Questions

1. Complete a copy of Table 62b on the next page using your experimental results and your data book. Then decide on the structure of these chlorides and the bonding found in them and fill in the last part of your table.

2. In the experiment you discovered that some of the chlorides are hydrolysed by water. Look up the equations for these reactions in your text-book(s). In the case of S$_2$Cl$_2$ you may find that most books do not give an equation for its hydrolysis. This is because a mixture of products is obtained. We suggest the following equation:

$$2S_2Cl_2(l) + 2H_2O(l) \rightarrow SO_2(aq) + 3S(s) + 4HCl(aq)$$

Table 62b Properties of Chlorides of Period 3

Formula of chloride	NaCl	MgCl$_2$	AlCl$_3$	SiCl$_4$	PCl$_3$	S$_2$Cl$_2$	Cl$_2$
Melting-point/°C							
Boiling-point/°C							
Physical state at r.t.p.*							
$\Delta H_f^{\ominus}$/kJ mol^{-1}							
$\Delta H_f^{\ominus}$ per mole of Cl/kJ mol^{-1}							
Conductivity of liquid							
Action of water							
pH of aqueous solution							
Solubility in hexane							
Structure							
Bonding							

*r.t.p. = room temperature and pressure (i.e. 20 °C and 1 atm)

147

EXPERIMENT 63
Preparing anhydrous aluminium chloride

Aim

In this experiment you gain experience in setting up an assembly of glassware in order to carry out an inorganic synthesis. You also calculate the percentage yield of your product.

Introduction

You prepare chlorine by adding concentrated hydrochloric acid to potassium manganate(VII), dry it using anhydrous calcium chloride and then pass it over heated aluminium foil in a combustion tube. When the reaction is complete you weigh your collected product and calculate the percentage yield.

Requirements

safety spectacles and gloves
access to fume cupboard
forceps
glass rod
calcium chloride, anhydrous, $CaCl_2$
long spatula
combustion tube
ceramic wool
bung fitted with a short piece of glass tubing
access to a balance capable of weighing to within 0.01 g
aluminium foil, Al
absorption tube
soda-lime
receiver bottle with two holed bung (connected to absorption tube and
  combustion tube)
ruler
rubber tubing for connections
3 clamps, bosses and stands
potassium manganate(VII), $KMnO_4$ — — — — — — — — — — — — — — — — — — —
pear-shaped flask with ground-glass joint, 50 cm^3
ground-glass adapter with T-connection
cylindrical funnel with ground-glass joint
hydrochloric acid, concentrated, HCl — — — — — — — — — — — — — — — —
Bunsen burner
specimen tube with lid
labels
access to a desiccator

Procedure

There are several steps in assembling the apparatus for this experiment. The diagram below (Fig. 38) gives you an idea of what you are aiming for. Details of the separate stages are given at appropriate points in the procedure.

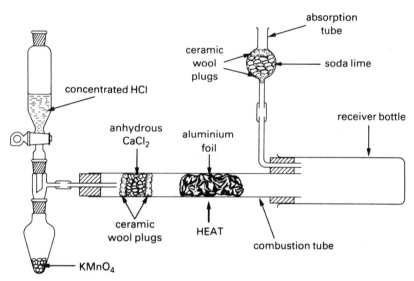

Fig. 38.   Preparation of aluminium chloride

1.   Using forceps and a glass rod, put some granular anhydrous calcium chloride between two loose plugs of ceramic wool near the entrance of the combustion tube. (Make sure that the calcium chloride fills the cross-section of the tube without preventing the free flow of chlorine.) Attach the bung fitted with glass tubing at the entrance of the combustion tube.

2.   Weigh about 0.25 g of aluminium foil, crumple it loosely, and put it in the combustion tube, as shown in Fig. 39.

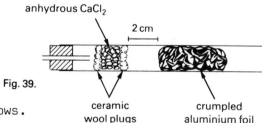

Fig. 39.

3.   Loosely pack the absorption tube as follows.

   (a)   Using forceps, push in a plug of ceramic wool

   (b)   Fill with soda-lime.

   (c)   Using forceps again, close with a second plug of ceramic wool. Both plugs must be loose to avoid blockage.

4. Fit the combustion tube and absorption tube to the receiver bottle as shown in Fig. 38. Clamp the apparatus so that the combustion tube is about 15 cm above the base of the fume cupboard.

5. Weigh about 5 g of potassium manganate(VII) and place it in the pear-shaped flask.

6. Fit the pear-shaped flask with the adapter and dropping funnel. Then clamp the flask and connect it to the combustion tube, as shown in Fig. 38.

You are now ready to start the preparation.

7. Get your teacher to check the apparatus.

8. Make sure that the tap of the dropping funnel is closed.

9. Pour 10 cm³ of concentrated hydrochloric acid into the dropping funnel.

10. Allow a few drops of acid to trickle on to the potassium manganate(VII) and allow chlorine to displace air from the apparatus. Then let the acid continue dripping slowly into the flask.

11. Heat the aluminium gently near the calcium chloride until a bright glow shows that the chlorine is reacting exothermically with the aluminium. If the glow is very bright, remove the heat till it subsides.

12. Continue heating, all around the tube, moving the flame slowly towards the receiver bottle, until reaction is complete.

13. When reaction is complete, stop the trickle of acid and let the combustion tube cool (about 10 minutes). Meanwhile weigh the empty specimen tube and top.

14. Remove the receiver bottle and, using your spatula, quickly scrape the product into the specimen tube. Place the top on the specimen tube and weigh it.

15. Record your results in a copy of Results Table 63.

16. Label the specimen tube with your name and the name of the product and store it in a desiccator.

Results Table 63

Mass of aluminium	g
Mass of empty specimen tube, m_1	g
Mass of specimen tube and product, m_2	g
Mass of product, $m = (m_2 - m_1)$	g
% yield	

Questions

1. Why is it important to use dry chlorine in this experiment?

2. What impurity would be present in the product if damp chlorine were used? Write an equation for the reaction giving this impurity.

3. Calculate and comment on the % yield of your product.

4. Why is your product stored in a desiccator?

EXPERIMENT 64

Investigating the properties
of Period 3 oxides

Aim

The purpose of this experiment is to examine
the oxides of Period 3 elements and describe
their structure and bonding.

Introduction

You carry out an investigation along similar lines to the work you did on
the chlorides of the elements in Period 3 (Experiment 63). However, you
will not be asked to test the oxides with hexane because, unlike the
covalent chlorides, most of the oxides are not composed of discrete
molecules. Therefore, they are unlikely to dissolve in hexane and simple
experiments cannot distinguish between insolubility and slight solubility.

Requirements

safety spectacles
access to a fume cupboard
6 test-tubes
test-tube rack
1 measuring cylinder, 10 cm^3
1 measuring cylinder, 100 cm^3
distilled water
thermometer, 0-100 °C
1 spatula
universal indicator solution and colour chart
teat-pipette
pH paper
sodium peroxide, Na_2O_2 —
splints
magnesium oxide, MgO
aluminium oxide, Al_2O_3
phosphorus(V) oxide, P_4O_{10} —
silicon(IV) oxide, SiO_2
access to sulphur dioxide cylinder or generator, SO_2 — — — — — — — —
Drechsel bottle
glass tubing with right-angled bend
rubber tubing for connections.

Hazard Warning

Phosphorus(V) oxide is corrosive and irritates eyes, skin and lungs.
Sodium peroxide is also corrosive and a powerful oxidant. Sulphur
dioxide is a toxic gas with a choking smell. Therefore you must:

DO THE EXPERIMENT AT THE FUME CUPBOARD
WEAR SAFETY SPECTACLES
AVOID CONTACT WITH SKIN

Procedure

A. Appearance

Examine your oxide samples, and in a larger copy of Results Table 64a note for each:

(a) whether it is solid, liquid or gaseous,

(b) its colour (if any).

B. On mixing with water

1. Set up six test-tubes, side by side.

2. Into each tube pour about 5 cm³ of distilled water.

3. In the first test-tube place a thermometer.

    (a) Note the temperature.

    (b) Add half a spatula-tip of sodium peroxide and stir carefully with the thermometer.

    (c) Note after about one minute, (i) the temperature, (ii) whether the solid has dissolved and (iii) anything else you see. For example, is gas evolved at any time? If so, is the gas acidic? Can you identify it using a simple test?

    (d) Add 2-4 drops of universal indicator solution, compare the colour with the chart provided, and note the pH indicated, or use a piece of pH paper.

4. Repeat the above steps 3.(a)-(d) using, in turn, magnesium oxide, aluminium oxide, silicon(IV) oxide and phosphorus(V) oxide.

5. Measure the pH of the water in the sixth test-tube by adding 2-4 drops of universal indicator solution for comparison with the above.

6. Bubble sulphur dioxide slowly through the water in the sixth test-tube until there is no further change in the colour of the indicator. Note the final pH of the solution. (You will probably be given sulphur dioxide in liquid form in a cylinder. To obtain the gas you carefully open the valve and the sudden decrease in pressure inside the cylinder causes the surface liquid to vaporize. Make sure there is a Drechsel bottle between the cylinder and the water in case of suck-back. Alternatively, your teacher may suggest other ways of generating the gas.)

7. To test the solubility of sulphur dioxide lower the delivery tube from your generator to the bottom of the 100 cm³ measuring cylinder filled with water. Pass a slow steady stream of gas through the water and when the air has been expelled from your apparatus look for a change in the size of the sulphur dioxide gas bubbles as they rise up through the water.

Results Table 64a

	Na_2O_2	MgO	Al_2O_3	SiO_2	P_4O_{10}	SO_2
Appearance						
On mixing with water Initial temperature Final temperature Does it dissolve? pH of solution Other observation(s) (if any)						

Questions

1. Use your experimental results, your data book and your text-book(s) if necessary to complete a larger copy of Table 64b.

Table 64b

Formula of oxide	$Na_2O_2^*$	MgO	Al_2O_3	SiO_2	$P_4O_{10}^*$	SO_2^*	Cl_2O^*
Melting-point/°C							
Boiling-point/°C							
State at s.t.p.							
Action of water							
pH of aq. solution							
Acid/base nature							
Conductivity of liquid							
Solubility in hexane							
Structure							
Bonding							

Substances marked * represent the most familiar or readily available oxides. In general, the other oxides of that element have similar properties.

2. Write equations for the oxide-water reactions which took place.

3. Comment on the change in structure and bonding in the Period 3 oxides.

4. How does the acid-base nature of these oxides change across the period?

5. Can you relate this change to the change in structure and bonding that takes place along the period?

EXPERIMENT 65

The reactions of tin and lead
and their aqueous ions

Aim

The purpose of this experiment is two fold: to
show the reactions of tin and lead with acids
and to familiarise you with some of the common
reactions of Pb^{2+}(aq) and Sn^{2+}(aq).

Introduction

In this experiment you find out if the metallic character of tin and lead
is evident from their reactions with acids. You treat the elements with two
acids; an oxidizing agent (nitric acid) and a non-oxidizing acid (hydrochloric acid). In each case you attempt to identify any gases evolved.

The reactions of the divalent ions are included here to demonstrate the
relative stability of the +2 state in tin and lead.

Requirements

safety spectacles and gloves
15 test-tubes
2 test-tube racks
test-tube holder
Bunsen burner and mat
lead (small pieces)
hydrochloric acid, dilute, 2 M HCl
hydrochloric acid, concentrated, HCl – – – – – – – – – – – – – – – – – – –
3 teat-pipettes
beaker, 250 cm^3
wood splints
universal indicator paper
nitric acid, concentrated, HNO$_3$ – – – – – – – – – – – – – – – – – – –
tin (small pieces)
sticky labels for test-tubes
0.1 M solution of Sn^{2+} ions (in dilute hydrochloric acid)
0.1 M solution of Pb^{2+} ions –
sodium hydroxide solution, 2 M NaOH
ammonia solution, 2 M NH$_3$
potassium manganate(VII) solution, 0.01 M KMnO$_4$ (in dilute ethanoic acid)
potassium chromate(VI) solution, 0.1 M K$_2$CrO$_4$
sodium sulphide solution, 0.02 M Na$_2$S – – – – – – – – – – – – – –
potassium iodide solution, 0.1 M KI

Hazard warning

Lead compounds are harmful if ingested or absorbed through the skin.

Sodium sulphide is toxic and corrosive and evolves highly poisonous hydrogen sulphide gas on contact with acids.

Concentrated hydrochloric acid is very corrosive.

Concentrated nitric acid is very corrosive and a powerful oxidant.

Therefore you MUST:

WEAR PROTECTIVE GLOVES AND SAFETY SPECTACLES.

USE SODIUM SULPHIDE IN THE FUME CUPBOARD.

Procedure

1. Place a small piece of lead in each of 3 test-tubes.

2. To one test-tube add about 2 cm³ of dilute hydrochloric acid and heat gently. Can you see or detect a gas? If not, repeat the experiment carefully using about 2 cm³ of concentrated hydrochloric acid.

3. In the remaining test-tube add about 2 cm³ of concentrated nitric acid to the lead and heat gently.

4. Place a small piece of tin in each of two test-tubes and repeat steps 2 and 3.

5. Record your results in a larger copy of Results Table 65a.

6. Add approximately 2 cm³ of the Sn^{2+} solution to each of seven test-tubes.

7. Add approximately 2 cm³ of the Pb^{2+} solution to each of seven test-tubes.

8. Add each of the following reagents to separate portions of the Sn^{2+} and Pb^{2+} solutions:

   (a) dilute sodium hydroxide solution, initially drop-by-drop, and then to excess;

   (b) ammonia solution, initially drop by drop, and then to excess;

   (c) about 2 cm³ of dilute hydrochloric acid, heat the mixtures and then cool them under running cold water;

   (d) about 2 cm³ of acidified potassium manganate(VII) solution;

   (e) about 1 cm³ of potassium chromate(VI) solution;

   (f) 5 drops of sodium sulphide solution (do this in the fume cupboard and dispose of the mixture by pouring into the fume cupboard sink);

   (g) about 2 cm³ of aqueous potassium iodide.

9. Record your observations in a larger copy of Results Table 65b.

Results Table 65a

Acid	Tin	Lead
Dilute hydrochloric acid		
Concentrated hydrochloric acid		
Concentrated nitric acid		

Results Table 65b

Reagent	$Sn^{2+}(aq)$(acidified)	$Pb^{2+}(aq)$
(a) Sodium hydroxide solution		
(b) Ammonia solution		
(c) Dilute hydrochloric acid		
(d) Acidified potassium manganate(VII) solution		
(e) Potassium chromate(VI) solution		
(f) Sodium sulphide solution		
(g) Potassium iodide solution		

Questions

1. (a) Complete the following equations:

   $Pb(s) + HCl(aq) \rightarrow$

   $Sn(s) + HCl(aq) \rightarrow$

2. Are the reactions between the elements and hydrochloric acid typical of metals? Explain your answer.

3. Reactions with nitric acid tend to be complex and equations are not generally required. The questions illustrate some general points.

   (a) Which gas did you detect when nitric acid reacted with tin and lead?

   (b) Do other metals behave in a similar way with nitric acid? Give one example.

(c) Why does nitric acid behave differently from hydrochloric acid?

4. Use your textbook(s) to complete and balance the equations in (a larger copy of) Table 65c. In the comments column you should describe the type of chemical reaction occurring and any other important feature(s).

Table 65c

Equations	Comments
$Sn^{2+}(aq) + OH^-(aq) \rightarrow$ $Sn(OH)_2(s) + OH^-(aq) \rightarrow$	The precipitate dissolves in excess NaOH to form a stannate(II) ion*, $Sn(OH)_6^{4-}$.
$Pb^{2+}(aq) + OH^-(aq) \rightarrow$ $Pb(OH)_2(s) + OH^-(aq) \rightarrow$	
$2MnO_4^-(aq) + 16H^+(aq) + 5Sn^{2+}(aq) \rightarrow$	
$Pb^{2+}(aq) + 2Cl^-(aq) \rightarrow$	Simple precipitation reaction. Precipitate soluble in hot water.
$Cr_2O_7^{2-}(aq)** + H^+(aq) + Sn^{2+}(aq) \rightarrow$	
$Pb^{2+}(aq) + CrO_4^{2-}(aq) \rightarrow$	
$Pb^{2+}(aq) + S^{2-}(aq) \rightarrow$	
$Sn^{2+}(aq) + S^{2-}(aq) \rightarrow$	
$Pb^{2+}(aq) + I^-(aq) \rightarrow$	

*Various formulae have been proposed for the stannate(II) ion ranging from SnO_2^{2-} for the anhydrous form to $Sn(OH)_4^{2-}$ and $Sn(OH)_6^{4-}$ for the hydrated forms. $Sn(OH)_6^{4-}$ seems the most probable. Similar variations have been proposed for the plumbate(II) ion $Pb(OH)_6^{4-}$.

**Chromate(VI) (CrO_4^{2-}) changes to dichromate(VI) $(Cr_2O_7^{2-})$ when acidified $(2CrO_4^{2-}(aq) + 2H^+(aq) \rightleftharpoons Cr_2O_7^{2-}(aq) + H_2O(l))$

5. In the experiment, potassium manganate(VII) solution was acidified with dilute ethanoic acid and not, as is usual, dilute hydrochloric acid or dilute sulphuric acid. With the aid of your text-book(s), explain why you think this change was made. Give any relevant equations in your answer and state what you would observe if the MnO_4^- solution were acidified with HCl(aq) or $H_2SO_4(aq)$.

6. State which of the divalent ions (Sn^{2+} and Pb^{2+}) is more stable. Explain your answer.

7. Predict the reaction between aqueous tin(II) ions and a solution of mercury(II) chloride. What do you think you would observe in this reaction?

157

EXPERIMENT 66

The preparations and reactions of tin(IV) oxide and lead(IV) oxide

Aim

The purpose of this experiment is to prepare tin(IV) oxide and lead(IV) oxide and compare their reactions and relative stabilities.

Introduction

In parts A and B of this experiment you react tin and lead with concentrated nitric acid. However, although this acid oxidizes tin to tin(IV) oxide, it oxidizes lead only to the +2 state. In order to reach the +4 state, you add a stronger oxidizing agent to the lead(II) compound (e.g. sodium chlorate(I) solution (sodium hypochlorite) in alkaline solution) to form lead(IV) oxide.

Since the preparation of lead(IV) oxide involves more than one stage we ask you to calculate the percentage yield in order to test the efficiency of the method and your practical techniques.

In part C of this experiment you compare the effect of heat on tin(IV) oxide and lead(IV) oxide and the effect of various reagents on these two oxides.

Hazard warning

Concentrated hydrochloric acid, nitric acid and sodium hydroxide are very corrosive. The vapours of the concentrated acids are also harmful to eyes, lung and skin. Concentrated nitric acid is an oxidizing agent. Therefore you MUST:

WEAR SAFETY SPECTACLES

AVOID CONTACT WITH SKIN

USE THESE SUBSTANCES IN THE FUME CUPBOARD

Nitrogen dioxide and chlorine are toxic. Therefore you MUST:

USE A FUME CUPBOARD WHEN THESE GASES ARE GENERATED.

Propanone is very flammable. Therefore you MUST:

KEEP THE STOPPER ON THE BOTTLE AS MUCH AS POSSIBLE
AND KEEP THE LIQUID AWAY FROM FLAMES.

Requirements

safety spectacles
access to fume cupboard and balance
disposable gloves
granulated tin
evaporating basin
Bunsen burner, tripod and gauze
2 measuring cylinders, 10 cm³ and 100 cm³
nitric acid, concentrated, HNO_3 — — — — — — — — — — — — — — — — —
3 teat pipettes
glass rod
spatula
suction filtration apparatus
distilled water
2 specimen tubes and sticky labels
lead foil
conical flask, 250 cm³
sodium hydroxide solution, 2 M NaOH — — — — — — — — — — — — — — — —
sodium chlorate(I) solution (hypochlorite), 1.5 M NaClO — — — — — — — —
nitric acid, 2 M HNO_3
propanone, CH_3COCH_3 —
tin(IV) oxide, SnO_2
lead(IV) oxide, PbO_2
10 test-tubes
test-tube holder
splints
hydrochloric acid, dilute, 2 M HCl
hydrochloric acid, concentrated, HCl — — — — — — — — — — — — — — — — —
blue litmus paper
sodium hydroxide solution, 8 M NaOH — — — — — — — — — — — — — — — —
potassium iodide solution (acidified), 1 M KI

Procedure

A. Preparation of tin(IV) oxide

1. Weigh about 1.5 g of granulated tin and place it in an evaporating
    basin in the fume cupboard.

2. Carefully pour 5 cm³ of concentrated nitric acid over the tin.
    Warm the mixture, if necessary, until the reaction is proceeding
    moderately.

3. As the evolution of gas subsides stir the contents of the basin
    carefully with a glass rod.

4. When the reaction appears to be complete add a few drops of concentrated
    nitric acid and warm carefully. If brown fumes are evolved, the reaction
    is not complete, in which case add a few more drops of concentrated
    nitric acid and warm until the brown fumes cease.

5. When cool, transfer the contents of the basin (mainly solid) to the
    suction filtration apparatus.

6. Wash the precipitate thoroughly with distilled water until there is no
    sign of a yellow colouration.

7. Transfer the precipitate (hydrated tin(IV) oxide) to an evaporating basin
    and heat (gently to start with, in order to avoid 'spitting', and then
    strongly).

8. When cool, record the appearance of the solid in a copy of Results
    Table 66a and transfer the solid to a labelled specimen bottle.

B. Preparation of lead(IV) oxide

1. Weigh about 5 g of lead (small pieces of foil), record the mass, and place it in a conical flask.

2. Pour 10 cm³ of distilled water into the flask followed by 10 cm³ of concentrated nitric acid.

3. Heat the flask gently in the fume cupboard until the reaction proceeds at a moderate rate.

4. Gradually add about 5-10 cm³ of distilled water to replace any water lost by evaporation.

5. When all the lead has reacted allow the mixture to cool to room temperature.

6. If crystals (of lead(II) nitrate) appear on cooling add a little distilled water to dissolve them.

7. Add dilute sodium hydroxide with a teat pipette and swirl the contents until a <u>permanent</u> faint white precipitate appears.

8. Pour in a further 50 cm³ of sodium hydroxide solution followed by 30 cm³ of sodium chlorate(I) solution (sodium hypochlorite) and carefully bring the contents of the flask to the boil.

9. Allow the precipitate to settle and carefully pour off the clear liquid.

10. Add 100 cm³ of distilled water, shake and again pour off the clear liquid.

11. Add 20 cm³ of dilute nitric acid followed by 50 cm³ of distilled water and shake again.

12. Filter the precipitate at the pump and wash thoroughly with distilled water, followed by two 5 cm³ portions of propanone.

13. Draw air through the solid for about 5-10 minutes to allow the propanone to evaporate.

14. Leave the solid in a warm place for a day or two and, when you are satisfied that it is completely dry, weigh it.

15. Record your results in a copy of Results Table 66a and calculate the percentage yield based on the lead used.

C. Reactions of tin(IV) oxide and lead(IV) oxide

Carry out the following tests on separate samples of the oxides which you have already prepared, and record your results in a larger copy of Results Table 66b. We suggest that you also use tin(IV) oxide prepared by the manufacturers for the tests with the concentrated acids and alkalis since the oxide prepared by the above method is particularly insoluble.

1. Heat a spatula-ful of each oxide separately in a test-tube. Identify any gases given off.

2. To half a spatula measure of each oxide (approximately 0.1 g) add about 1 cm³ of dilute hydrochloric acid in the fume cupboard. Identify any gas given off.

3. (a) Add approximately 0.1 g of lead(IV) oxide to 1 cm³ of concentrated hydrochloric acid in the fume cupboard, identify the gas evolved and note the colour of the final solution.

    (b) Add half a spatula-tip (or less) of tin(IV) oxide to 1 cm³ of concentrated hydrochloric acid and very cautiously warm the mixture in the fume cupboard. Does any of the solid dissolve?

160

(c)   Add half a spatula-tip (or less) of each oxide separately to 1 cm³ of concentrated sodium hydroxide (8 M NaOH) and very cautiously warm the mixture in the fume cupboard.

(d)   Add approximately 0.1 g of each oxide separately to 2 cm³ of acidified KI solution.

Results Table 66a

Mass of Pb		Appearance of PbO_2	
Mass of PbO_2		Appearance of SnO_2	
% yield			

Results Table 66b

Test	Tin(IV) oxide	Lead(IV) oxide
(a) Heat		
(b) Dilute hydrochloric acid		
(c) Concentrated hydrochloric acid		
(d) Concentrated sodium hydroxide		
(e) Acidified potassium iodide		

Questions

1.   Write an equation for the effect of heat on lead(IV) oxide.

2.   What is happening in the reactions between (a) lead(IV) oxide and dilute hydrochloric acid and (b) lead(IV) oxide and acidified potassium iodide?

3.   What is the yellow liquid formed when lead(IV) oxide reacts with concentrated hydrochloric acid?

4.   Write equations for the reactions between (a) tin(IV) oxide and concentrated sodium hydroxide and (b) lead(IV) oxide and concentrated sodium hydroxide.

5.   What is the acid-base nature of tin(IV) oxide and lead(IV) oxide? (Bear in mind that concentrated sulphuric acid will attack tin(IV) oxide to form tin(IV) sulphate.)

6.   Dilead(II) lead(IV) oxide (red lead), Pb_3O_4, behaves in many respects as a mixture of lead(II) oxide and lead(IV) oxide and the formula $Pb_2^{II}Pb^{IV}O_4$ (or $2PbO \cdot PbO_2$) is often recommended. Predict the effect of heat on the substance and write the chemical equation.

7.   Predict the effect of heating separate samples of tin(II) oxide and lead(II) oxide. Find out what actually happens and write equations.

EXPERIMENT 67

Observation and deduction exercise

Aim

The purpose of this experiment is to give you further practice in the investigation of unknown substances.

Introduction

The procedure which follows is taken from two separate A-level practical examination papers. Since the format of each paper is slightly different we have divided this experiment into two sections. In the first section you carry out tests on a powder Q, and in the second section you test a solution of a compound H and also a solid I.

Requirements

safety spectacles
10 test-tubes in rack
test-tube holder
sample of powder Q —
spatula
Bunsen burner and mat
wood splints and litmus papers
wash-bottle of distilled water
funnel and filter paper
nitric acid, dilute, 2 M HNO_3
potassium iodide solution, 0.1 M KI
sodium hydroxide solution, 2 M NaOH
solution of substance H
silver nitrate solution, 0.02 M $AgNO_3$
iron(III) chloride solution (neutralized), 0.3 M $FeCl_3$
potassium thiocyanate solution, 0.5 M KSCN
mercury(II) chloride solution, 0.1 M $HgCl_2$ — — — — — — — — — — — — —
sample of solid I
1 boiling-tube
sulphuric acid, dilute, 2 M H_2SO_4
potassium chromate(VI) solution, 0.5 M K_2CrO_4
other chemicals are available from your teacher.

Hazard warning

Mercury(II) chloride solution is poisonous. Therefore you MUST:

AVOID CONTACT WITH SKIN AND DISPOSE OF RESIDUES CONTAINING MERCURY BY POURING INTO THE FUME CUPBOARD SINK WITH PLENTY OF RUNNING COLD WATER

Procedure

A. Tests on Q. Test the powder Q as follows:

1. Heat a portion in a test-tube until reaction ceases. Test any gas evolved.

2. Warm a portion with dilute nitric acid. Filter, if necessary. Retain the solution or filtrate for (3).

3. Test portions of the solution or filtrate from (2) with

    (a) aqueous potassium iodide,

    (b) aqueous sodium hydroxide.

    Carefully observe what happens, and report fully. What tentative inferences do you draw from these experiments? Carry out and report on one further experiment which tests your inferences. This experiment can be made on Q or on the products of the above reactions. Full credit will not be given unless your answer discloses the method (including the scale of your experiments), careful observations, and some comment on the type of chemical reactions involved in the experiments.

The record of your work must be made in (larger copies of) the tables provided.

Results Table 67a Tests with Q

Test	Method	Observations	Inferences
(1) Heat			
(2) Dilute nitric acid			
(3) Test portions of solution or filtrate from (2) with (a) potassium iodide solution, (b) sodium hydroxide solution.			

Results Table 67b Experiment to test inference

Inference	Test and Observations	Conclusion

B. Tests on H and I

   You are provided with a solution of a compound H and a solid I. Carry out the following tests and record your observations and inferences in a larger copy of Results Table 67c. Then answer the question which follows the table.

Results Table 67c

Test	Observations	Inferences
1. To 1 cm³ of the solution of H add aqueous silver nitrate followed by dilute nitric acid.		
2. To 1 cm³ of the solution of H add aqueous sodium hydroxide until in excess.		
3. To 1 cm³ of aqueous iron(III) chloride add a few drops of aqueous potassium thiocyanate. To this solution add some of the solution of H.		
4. To 1 cm³ of aqueous mercury(II) chloride add a little of the solution of H, then excess.		
5. Heat some of I in a pyrex boiling-tube. Allow to cool. Add 8-10 cm³ of dilute nitric acid to the residue and boil the mixture for 1 or 2 minutes. Filter if necessary and use portions of the cool solution for the following tests: (a) To 1 cm³ of the solution add aqueous sodium hydroxide. (b) To 1 cm³ of the solution add dilute sulphuric acid. (c) To 1 cm³ of the solution add aqueous potassium chromate(VI) (potassium chromate).		

Now answer the following question:

Comment on the nature of the metal contained in H and on the oxidation states of this metal in the compounds involved in reactions 2 to 4.

EXPERIMENT 68
Illustrating the oxidation states of vanadium

Aim

The purpose of this experiment is to illustrate the presence of several different
oxidation states for vanadium, and to show
how it is possible to change from one
oxidation state to another.

Introduction

Starting with a solution containing vanadium(V) in acid conditions, you use
powdered zinc as a reducing agent. Colour changes indicate the formation of
other oxidation states. After this first series of reactions you perform
some changes between oxidation states by using a variety of oxidizing and
reducing agents.

Requirements

safety spectacles and gloves
conical flask, 100 cm³
spatula
ammonium polytrioxovanadate(V), NH_4VO_3 (ammonium metavanadate)
measuring cylinder, 25 cm³
sulphuric acid, dilute, 1 M H_2SO_4
sulphuric acid, concentrated, H_2SO_4 — — — — — — — — — — — — — —
5 test-tubes and holder
test-tube rack
zinc dust, Zn
Bunsen burner, tripod, gauze and bench mat
filter funnel and paper
potassium manganate(VII) solution, 0.02 M $KMnO_4$ (permanganate)
sodium sulphite, Na_2SO_3
potassium iodide solution, 0.05 M KI
sodium thiosulphate solution, 0.1 M $Na_2S_2O_3$
wash-bottle of distilled water

Hazard warning

Concentrated sulphuric acid is corrosive and reacts
violently with water. Therefore you MUST

WEAR SAFETY SPECTACLES AND GLOVES.
WHEN DILUTING, ADD ACID TO WATER, <u>NOT</u> WATER TO ACID.
MOP UP SMALL SPILLAGES WITH EXCESS WATER.

Procedure

First complete a copy of Table 68a to assist you in identifying the different oxidation states of vanadium.

Table 68a

Ion (hydrated)	*VO_2^+	VO^{2+}	V^{3+}	V^{2+}
Colour	yellow	blue	green	violet
Oxidation state				
Name				

*The VO_3^- ion in the ammonium salt is converted to VO_2^+ by acid:

$$VO_3^-(aq) + 2H^+(aq) \rightleftharpoons VO_2^+(aq) + H_2O(l)$$

1. Place about 0.25 g (one spatula measure) of ammonium trioxovanadate(V) in a conical flask and add about 25 cm³ of dilute sulphuric acid. Carefully add about 5 cm³ of concentrated sulphuric acid and swirl the flask until you obtain a clear yellow solution.

2. Pour about 2 cm³ of this vanadium(V) solution into each of two test-tubes ready for later tests.

3. To the conical flask add 1 - 2 g (one spatula measure) of zinc dust, a little at a time. Swirl the flask at intervals and record any observed colour changes in a copy of Results Table 68b.

4. When the solution has become violet (you may need to heat the flask for this final change), filter about 2 cm³ into each of three test-tubes.

5. To one of the three tubes add, a little at a time, an excess of acidified potassium manganate(VII) solution, shaking after each addition, until no further change is observed.

6. Keep the other test-tubes of solutions for later tests. Answer the first two questions (next page) and complete Results Table 68b as far as you can before doing further tests.

7. To one of the tubes containing vanadium(V) add a little sodium sulphite and shake. Filter if cloudy. Now boil carefully (at a fume cupboard) to remove excess sulphur dioxide and add about the same volume of the vanadium(II) solution. Record your observations.

8. To the second of the tubes containing vanadium(V), add about 2 cm³ potassium iodide solution and mix. Then add about 2 cm³ of sodium thiosulphate solution. Record your observations.

9. Keep the last tube of vanadium(II) solution for a final experiment you may wish to do after answering question 7.

Results Table 68b

Test	Observations	Summary of reaction
Ammonium vanadate + acid	White solid dissolved to a yellow solution.	$\overset{+5}{V}O_3^- \rightarrow \overset{+5}{V}O_2^+$
Vanadium(V) + zinc		
Vanadium(II) + manganate(VII)		
Vanadium(V) + sulphite Add vanadium(II)		
Vanadium(V) + iodide + thiosulphate		
Vanadium(II) + - - - - -		

Questions

1. How do you explain the <u>first</u> appearance of a green colour in the solution?

2. What are the subsequent changes in colour and why do these changes occur?

You will need the following electrode potentials in order to answer some of the remaining questions:

$$Zn^{2+}(aq) + 2e^- \rightleftharpoons Zn(s) \qquad\qquad E^\ominus = -0.76 \text{ V}$$

$$V^{3+}(aq) + e^- \rightleftharpoons V^{2+}(aq) \qquad\qquad E^\ominus = -0.26 \text{ V}$$

$$VO^{2+}(aq) + 2H^+(aq) + e^- \rightleftharpoons V^{3+}(aq) + H_2O(l) \qquad E^\ominus = +0.34 \text{ V}$$

$$VO_2^+(aq) + 2H^+(aq) + e^- \rightleftharpoons VO^{2+}(aq) + H_2O(l) \qquad E^\ominus = +1.00 \text{ V}$$

$$SO_4^{2-}(aq) + 4H^+(aq) + 2e^- \rightleftharpoons H_2SO_3(aq) + H_2O(l) \qquad E^\ominus = +0.17 \text{ V}$$

$$I_2(aq) + 2e^- \rightleftharpoons 2I^-(aq) \qquad\qquad E^\ominus = +0.54 \text{ V}$$

3. What did you observe when you added iodide ions to vanadium(V)? What caused this colour?

4. Why did you add sodium thiosulphate?

5. Why does reduction with iodide not give the same result as reduction with zinc?

6. What did you observe when you added sulphite ions to acidified vanadium(V) solution? Does this result correspond with a prediction made using the $E^\ominus$ values? (Hint: sulphite ions and acid react to give what?)

7. How would you set about finding a suitable oxidizing agent for the oxidation of vanadium(II) to vanadium(III) and no further? Does one appear in the table above?

EXPERIMENT 69
Illustrating the oxidation
states of manganese

Aim and Introduction

The purpose of this experiment is to make
samples of some of the less common
oxidation states of manganese, using
methods predicted in the next three
exercises.

Exercise 1 Predict the feasibility of making Mn(VI) from Mn(VII) and
Mn(IV) under conditions of varying pH. Use these
electrode potentials:

$2MnO_4^-(aq) + 2e^- \rightleftharpoons 2MnO_4^{2-}(aq);$ $E^\ominus = 0.56$ V

$MnO_4^{2-}(aq) + 4H^+(aq) + 2e^- \rightleftharpoons MnO_2(s) + 2H_2O(l);$ $E^\ominus = 2.26$ V

$MnO_4^{2-}(aq) + 2H_2O(l) + 2e^- \rightleftharpoons MnO_2(s) + 4OH^-(aq);$ $E^\ominus = 0.60$ V

Exercise 2 Predict the feasibility of making Mn(III) from Mn(II) and
Mn(IV) under conditions of varying pH. Use these
electrode potentials:

$MnO_2(s) + 4H^+(aq) + e^- \rightleftharpoons Mn^{3+}(aq) + 2H_2O(l);$ $E^\ominus = 0.95$ V

$Mn^{3+}(aq) + e^- \rightleftharpoons Mn^{2+}(aq);$ $E^\ominus = 1.51$ V

$Mn(OH)_3(s) + e^- \rightleftharpoons Mn(OH)_2(s) + OH^-(aq);$ $E^\ominus = -0.10$ V

$MnO_2(s) + 2H_2O(l) + e^- \rightleftharpoons Mn(OH)_3 + OH^-(aq);$ $E^\ominus = 0.20$ V

Exercise 3 Predict the feasibility of making Mn(III) from Mn(II) and
Mn(VII) in acid conditions. The relevant half-cell
potentials are:

$Mn^{3+}(aq) + e^- \rightleftharpoons Mn^{2+}(aq);$ $E^\ominus = 1.51$ V

$MnO_4^-(aq) + 8H^+(aq) + 5e^- \rightleftharpoons Mn^{2+}(aq) + 4H_2O(l);$ $E^\ominus = 1.50$ V

Requirements

safety spectacles and protective gloves
6 test-tubes and rack
potassium manganate(VII) (permanganate) solution, 0.01 M KMnO$_4$
sulphuric acid, dilute, 1 M H$_2$SO$_4$
sodium hydroxide solution, 2 M NaOH $-$ $-$ $-$ $-$ $-$ $-$ $-$ $-$ $-$ $-$ $-$ $-$ $-$ $-$ $-$
manganese(IV) oxide (manganese dioxide), MnO$_2$
spatula
stirring rod
filter funnel and 3 papers
manganese(II) sulphate-4-water, MnSO$_4$·4H$_2$O
sulphuric acid, concentrated, H$_2$SO$_4$ $-$ $-$ $-$ $-$ $-$ $-$ $-$ $-$ $-$ $-$ $-$ $-$ $-$ $-$
spatula and stirring rod
wash-bottle of distilled water

Hazard warning

Concentrated sulphuric acid is corrosive and reacts
violently with water. Therefore you MUST:

WEAR SAFETY SPECTACLES AND GLOVES.
WHEN DILUTING, ADD ACID TO WATER, NOT WATER TO ACID.
MOP UP SMALL SPILLAGES WITH EXCESS WATER.

Procedure. Part A   Making Mn(VI) from Mn(VII) and Mn(IV)

Refer to Exercise 1 on the previous page in which you predicted that
it should be possible to make Mn(VI) from Mn(VII) and Mn(IV).

1. Put about 5 cm^3 of potassium manganate(VII) solution in each of three
    test-tubes.

2. To one of the three tubes add about 3 cm^3 dilute sulphuric acid and to
    another add about 3 cm^3 sodium hydroxide solution.

3. To each of the three tubes add a little solid manganese(IV) oxide and
    stir for about a minute.

4. Filter enough of each mixture into a clean tube to see the colour of
    the filtrate clearly. Use a fresh filter paper for each mixture.

5. One of the tubes should now have in it a clear green solution of Mn(VI).
    Add to this a little dilute sulphuric acid.

Questions. Part A.

1. Explain why only one of the three mixtures reacted to give green Mn(VI).

2. What happened when acid was added to Mn(VI)? Explain.

Procedure. Part B   Making Mn(III) from Mn(II) and Mn(VII)

6. Dissolve about 0.5 g manganese(II) sulphate in about 2 cm^3 of dilute
    sulphuric acid in a test-tube.

7. Carefully add about 10 drops concentrated sulphuric acid and
    cool the tube under a running tap.

8. Add a few drops of potassium manganate(VII) solution to obtain a deep
    red solution of Mn(III).

9. Dilute the red solution with about five times its volume of water, wait
    a few moments, and note any colour change.

Question. Part B.

3. Explain what happened when the Mn(III) solution was diluted.

Procedure. Part C Making Mn(III) from Mn(II) and Mn(IV)

10. In each of two test-tubes, dissolve a little manganese(II) sulphate in water and add an equal volume of sodium hydroxide solution to obtain a precipitate of manganese(II) hydroxide.

11. To one of the two tubes, add a little manganese(IV) oxide and stir.

12. Let both tubes stand for a few minutes, and note any changes.

Questions. Part C.

4. Can you see any sign of Mn(III) in the tubes?

5. What is different about the conditions of this experiment (part C) compared with the last (part B) which makes its success less likely?

6. What happens in the test-tube which had no manganese(IV) added? Suggest an explanation. (Hint: is the change greater in the upper part of the mixture?)

EXPERIMENT 70
Relative stabilities of some complex ions

Aim

The purpose of this experiment is to test predictions of ligand replacement reactions made using stability constants.

Introduction

In a series of test-tube reactions you examine a number of ligan replacement reactions. One type, in which charged and uncharged ligands are represented as lig⁻ and LIG respectively, can be represented:

$$Cu(LIG)_4{}^{2+} + 4\ lig^- \rightleftharpoons Cu(lig)_4{}^{2-} + 4\ LIG$$

You use stability constants to predict the outcome before mixing suitable pairs of solutions.

Requirements

6 test-tubes
test-tube rack
copper(II) sulphate solution, 0.20 M $CuSO_4$
wash-bottle of distilled water
sodium chloride solution, saturated NaCl
sodium ethanedioate solution, 0.20 M $(CO_2Na)_2$ (sodium oxalate)
1,2-diaminoethane solution, 0.10 M $(CH_2NH_2)_2$
edta solution (sodium salt), 0.10 M $C_{10}H_{14}O_8N_2Na_2$

Procedure

1. Complete the four prediction columns (headed 'P') in a copy of Results Table 70, using the stability constants given. Use a '√' to indicate 'replacement' and a 'x' to indicate 'no reaction'.

2. Place about 1 cm³ of copper sulphate solution in each of five test-tubes.

3. To the first tube add about 5 cm³ distilled water. This tube is for colour comparison with the others.

4. To the second tube add about 5 cm³ sodium chloride solution a little at a time, noting any colour change.

5. To the third tube add about 5 cm³ sodium ethanedioate in the same way.

6. Similarly, to the fourth and fifth tubes add diaminoethane and edta solutions respectively.

7. The colours in the five tubes are predominantly due to the five complex ions shown in Results Table 70. Write down the colours in your table, and fill in the first results column (headed 'R'). Again, use a '√' or an 'x' to show whether or not replacement of ligands has occurred.

Note that, in addition to the abbreviation edta, we use 'ox' to represent the ethanedioate (oxalate) ligand and 'en' to represent the 1,2-diaminoethane ligand (ethylenediamine)

Results Table 70

Ligand	H_2O	Cl^-	$C_2O_4^{2-}$ = ox	$NH_2C_2H_4NH_2$ = en	$C_{10}H_{14}O_8N_2^{4-}$ = edta
Complex ion	$Cu(H_2O)_4^{2+}$	$CuCl_4^{2-}$	$Cu(ox)_2^{2-}$	$Cu(en)_2^{2+}$	$Cu(edta)^{2-}$
Colour					
Stability constant	———	4.0 x 10 mol dm	2.1 x 10 mol dm	———	6.3 x 10 mol dm

Predictions (P) and results (R).   ✓ = replacement, x = none

Test	P	R	P	R	P	R	P	R	P	R
Add H_2O	▓	▓					▓			
Add Cl^-			▓	▓			▓			
Add ox					▓	▓	▓			
Add en							▓	▓		
Add edta							▓		▓	▓

8. Divide the solution containing $CuCl_4^{2-}$ ions into four parts. To each of these add an excess of one of the other four ligands in turn – use the original ligand solutions and not the complex ion solutions you have made. Complete the second results column according to whether or not you think replacement has occurred.

9. Repeat step 8 for the remaining complex ion solutions and complete the remaining results columns.

Questions

1. Could you have made any tentative predictions without knowing stability constants?

2. Estimate the approximate stability constant for $Cu(en)_2^{2+}$.

3. One of the ligand exchange reactions appeared to be readily reversible. Which one was this?

4. Calculate the ratio $[Cu(H_2O)_4^{2+}(aq)]/[CuCl_4^{2-}(aq)]$ for the following values of $[Cl^-(aq)]$:

   (a) 5.0 mol dm^{-3} (in saturated NaCl)

   (b) 0.050 mol dm^{-3} (in dilute NaCl).

   How do these ratios help to explain the reversible nature of the ligand exchange?

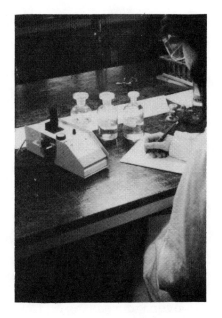

EXPERIMENT 71
Determining the formula of a complex ion

Aim

The purpose of this experiment is to use colorimetry to find the formula of the complex ion formed from copper(II) ions and ammonia. The same method can be used for many other complex ions.

Introduction

Using different proportions of copper(II) ion and ammonia you prepare mixtures and examine them in a colorimeter. The mixture which shows the most intense colour (and, therefore, the greatest absorbance) indicates the proportions of Cu^{2+} and NH_3 in the complex ion.

Requirements

safety spectacles
3 beakers, 100 cm³
3 burettes, 50 cm³ or 3 graduated pipettes, 10 cm³
3 small funnels for filling burettes
8 test-tubes, to fit colorimeter
8 corks or bungs to fit test-tubes
test-tube rack and 3 burette stands
ammonium sulphate solution, 2.0 M $(NH_4)_2SO_4$
copper(II) sulphate solution, 0.10 M $CuSO_4$
ammonia solution, 0.10 M NH_3
colorimeter, with filters
distilled water

Procedure

1. Prepare a different mixture in each of 8 test-tubes by running in from burettes or graduated pipettes the volumes of the three solutions shown in Results Table 71. Cork each tube and shake to mix thoroughly.

Results Table 71

Tube number	1	2	3	4	5	6	7	8
Volume of $(NH_4)_2SO_4$/cm³	15	5	5	5	5	5	5	5
Volume of $CuSO_4$/cm³	0.0	1.0	1.5	2.0	2.5	3.0	4.0	5.0
Volume of NH_3/cm³	0.0	9.0	8.5	8.0	7.5	7.0	6.0	5.0
Colorimeter reading	0.0							

2. Choose a suitable filter for this experiment. If you are not quite sure which filter to use, look at the notes at the end of this procedure section.

3. With the filter in position, put tube 1 into the colorimeter, cover it to exclude stray light, and turn the adjusting knob so that the meter shows zero absorbance (or, on some colorimeters, 100% transmission).

   Ideally, this should set the intensity of the light reaching the tubes for the whole experiment but, because simple colorimeters tend to give a light intensity which is not constant, you should repeat this step immediately before each reading you take.

4. Still with tube 1 in the colorimeter, check for constancy of meter reading with the tube rotated and moved to slightly different positions. Mark the tube so that you can replace it in such a way as to get the same reading each time.

   If necessary, change the tube until you are satisfied. (With some of the simplest colorimeters, it may not be possible for you to be too fussy!)

5. Immediately after setting zero with tube 1, replace it with tube 2 and take a reading of absorbance (or % transmission). To minimise errors due to irregular tube diameter and position, rotate the tube and take an average reading. (If you are lucky enough to be using matched tubes, put the mark in the same position each time.)

   Also, check the zero again after the reading and repeat if necessary. Record the average reading in a copy of Results Table 71.

6. Repeat steps 3 and 4 before taking further colorimeter readings for each of the remaining coloured solutions in turn.

7. Plot a graph of meter reading against volume of reactants as shown in Fig. 40. If your meter shows absorbance, the graph will show a peak; if it shows transmittance, the graph will show a trough.

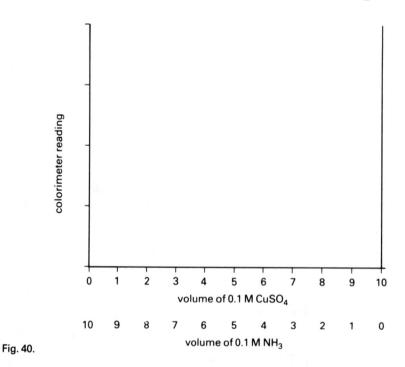

Fig. 40.

8. From the graph, determine the volumes of $CuSO_4$ and NH_3 corresponding to the peak (or trough if you are working with transmittance). These volumes may not necessarily be those used in any of the tubes, but should give a simple whole-number ratio.

9. Use the simple whole-number ratio obtained from the graph to write the formula for the complex ion.

Choosing a filter

Ideally, the filter lets through only light of the particular wavelength which is absorbed by the complex ion. So for a blue complex ion, which absorbs yellow light, you need a yellow filter - yellow is the complementary colour to blue. See Fig. 41.

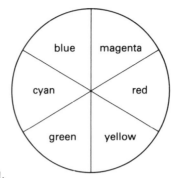

Fig. 41.

If you are not clear which filter has the complementary colour of the solution you are using, then follow the detailed procedure below.

1. Put any one of the available filters into the slot in the colorimeter, and insert tube 1, covering it to exclude stray light. Turn the adjusting knob so that the meter shows zero absorbance (or, on some colorimeters, 100% transmission). Mark the rim of the tube so that you can replace it in the same position.

2. Replace tube 1 with tube 5 and take a reading of absorbance (or of % transmission). Mark this tube too so that you can replace it in the same position.

3. Repeat steps 1 and 2 above for all the filters in turn.

4. Choose the filter which gives the greatest absorbance (or least transmission) and go back to step 3 of the procedure section.

Questions

1. What is the function of the ammonium sulphate in this experiment? If you do not know the answer, try adding ammonia solution to copper sulphate solution without any ammonium sulphate present. If you still need a clue, look up the common ion effect.

2. Why does the absorbance decrease after the peak while the amount of added copper ions increases?

3. Why does the absorbance increase up to the peak while the amount of added ammonia decreases?

4. How would your results be affected by using the wrong filter?

5. How would your results be affected by using no filter at all?

Extension

The method described above can be used with very little modification (if any) to find the formulae for many other complex ions. You may wish to try, for example, using a solution of diaminoethane (ethylene-diamine) in place of the ammonia in Experiment 71. If you do this, the ammonium sulphate is not necessary and should be replaced with water.

Other complexes you might try are those made from:

nickel(II) ions and edta

iron(III) ions and thiocyanate ions

Note that the volumes given in Results Table 71 may need adjusting for different complexes. Ask your teacher for advice.

EXPERIMENT 72
Some redox chemistry of copper

Aim and Introduction

The purpose of this experiment is to prepare two copper(I) compounds by reduction of copper(II). Different conditions are used in Parts A, B and C.

Requirements

safety spectacles
boiling-tube and holder
spatula
copper turnings, Cu
copper(II) chloride-6-water, $CuCl_2 \cdot 6H_2O$
sodium chloride, NaCl
wash-bottle of distilled water
Bunsen burner and bench mat
beaker, 250 cm^3
4 test-tubes
test-tube rack
sodium sulphite-7-water, $Na_2SO_3 \cdot 7H_2O$
copper(II) sulphate solution, 0.10 M $CuSO_4$
potassium iodide solution, 0.10 M KI
sodium thiosulphate-5-water, $Na_2S_2O_3 \cdot 5H_2O$

Procedure - Part A

1. Put 1 spatula measure of copper turnings, two of copper(II) chloride and two of sodium chloride in a boiling-tube and add about 10 cm^3 of distilled water.

2. Heat the tube till the contents <u>just</u> boil, and maintain this temperature, swirling the tube from time to time, for several minutes until the solution appears not to be darkening in colour any further.

3. Allow the tube to cool a little, and pour the solution into about 50 cm^3 of distilled water in a beaker.

4. Put the beaker on one side while you do Part B of the experiment and then re-examine it.

Procedure - Part B

1. Dissolve one spatula measure of copper(II) chloride in water in a test-tube about one-third full.

Procedure - Part B (continued)

2. Dissolve one spatula measure of sodium sulphite in water in another test-tube about one-third full.

3. Mix the two solutions and allow to stand for a few minutes.

4. If necessary, pour off the solution to examine the white solid.

Questions - Parts A and B

1. What is the white solid produced in each case, and why has it no colour?

2. What is the reducing agent in each case?

3. Why was heat required in Part A and not in Part B?

4. What is the function of the excess of chloride ions in Part A?

5. What causes the dark brown colour in Part A?

6. Why does the solution in contact with the white solid in Part A slowly turn blue on standing?

Procedure - Part C

1. Mix a little copper(II) sulphate solution in a test-tube with about twice the volume of potassium iodide solution and allow to settle.

2. Explain the change in appearance after step 2.

Questions - Part C

1. What do you observe in step 1?

2. Explain the change in appearance after step 2.

3. Why is the solution no longer blue after step 2?

4. What, therefore, is the creamy white solid?

Summarise the whole experiment in a larger copy of Results Table 72.

Results Table 72

	Method	Observations	Equation(s)
A. Preparation of			
B. Preparation of			
C. Preparation of			

EXPERIMENT 73
Investigating the use of
cobalt(II) ions as a catalyst

Aim

The purpose of this experiment is to test whether cobalt(II) ions will catalyse the oxidation of 2,3-dihydroxybutanedioate ions by hydrogen peroxide and to look for evidence of intermediate compound formation.

Introduction

In this short experiment, you simply mix the hot reactants and add the potential catalyst. If you see a coloured intermediate, you can attempt to stabilize it by cooling the mixture quickly.

Requirements

safety spectacles
beaker, 250 cm³
distilled water
spatula and stirring rod
potassium sodium 2,3-dihydroxybutanedioate, $CO_2K(CHOH)_2CO_2Na\cdot 4H_2O$
  (also known as potassium sodium tartrate or Rochelle salt)
Bunsen burner, tripod and gauze, bench mat
thermometer, 0 - 100 °C
measuring cylinder, 25 cm³
hydrogen peroxide solution, 1.7 M H_2O_2 — — — — — — — — — — — — — — — — —
2 test-tubes and rack
cobalt(II) chloride, $CoCl_2\cdot 6H_2O$
teat pipette

Procedure

1. Dissolve about 1 g of potassium sodium 2,3-dihydroxybutanedioate in about 50 cm³ water in a beaker and heat the solution to about 70 °C.

2. Add about 20 cm³ hydrogen peroxide solution, heat again to about 70 °C and then remove from heat. At this stage there should be little or no sign of reaction.

3. Dissolve about 0.25 g of cobalt(II) chloride in about 5 cm³ of water in a test-tube and add this to the hot solution. There will be an induction period before a reaction starts, so be patient! The length of the induction period decreases with increasing concentrations of reactants and catalyst and with increasing temperature. These can be varied if necessary.

4. As soon as the solution appears to be dark green, quickly transfer about 5-10 cm³ by teat pipette to a test-tube and cool it under the tap. If the green colour has disappeared before you have cooled the tube, repeat the experiment in such a way that the reaction proceeds more slowly.

Questions

1. The gas formed when the reaction starts has two components. What are these?

2. What is the dark green solution and why does the colour return to pink?

3. Why does cooling the test-tube preserve the green colour for a while?

EXPERIMENT 74

Catalysing the reaction between iodide ions
and peroxodisulphate ions

Aim

The purpose of this experiment is to find and
test suitable catalysts, from a range of
readily available transition metal ions, for
the reaction:

$$S_2O_8{}^{2-}(aq) + 2I^-(aq) \rightarrow 2SO_4{}^{2-}(aq) + I_2(aq)$$

Introduction

You can make predictions about the suitability of possible catalysts by
assuming that the mechanism of catalysis consists of two stages, either of
which can be the first:

    (i)   higher oxidation state of catalyst oxidizes I^-,

    (ii) lower oxidation state of catalyst reduces $S_2O_8{}^{2-}$.

You then perform a number of experiments, each with fixed starting concentrations of I^-, $S_2O_8{}^{2-}$ and possible catalyst. Thiosulphate ions and starch
are also added to enable you to measure the rate of reaction, and thus test
your predictions.

As the reaction proceeds, any iodine formed immediately reacts with the
thiosulphate:

$$I_2(aq) + 2S_2O_3{}^{2-}(aq) \rightarrow S_4O_6{}^{2-}(aq) + 2I^-(aq)$$

When all the thiosulphate has reacted, the free iodine gives a deep blue
colour with the starch. If t is the time taken for the blue colour to
appear, then $1/t$ is a measure of the initial rate of reaction.

Requirements

safety spectacles
4 test-tubes and rack
potassium iodide solution, 0.2 M KI
sulphuric acid, dilute, 1 M H_2SO_4
potassium manganate(VII) solution, 0.1 M $KMnO_4$
potassium chromate(VI) solution, 0.1 M K_2CrO_4
iron(III) chloride solution, 0.1 M $FeCl_3$
potassium peroxodisulphate solution 0.2 M $K_2S_2O_8$
manganese(II) sulphate solution, 0.1 M $MnSO_4$
chromium(III) chloride solution, 0.1 M $CrCl_3$
iron(II) sulphate solution, 0.1 M $FeSO_4$
4 burettes and stands   ⎫
4 beakers, 100 cm³   ⎬ (shared with other students)
4 small funnels   ⎭
starch solution, 0.2%
sodium thiosulphate solution, 0.01 M $Na_2S_2O_3$
conical flask, 150 cm³
boiling-tube
stopclock
wash-bottle of distilled water

Procedure

1. In a copy of Results Table 74 write the appropriate electrode potentials from your data book.

2. Use the electrode potentials to predict the feasibility of each step in the proposed mechanism. You should wait till these steps have been tested before predicting the overall feasibility.

Results Table 74

Transition element		Chromium	Manganese	Iron
Oxidation states used		Cr(VI) Cr(III)	Mn(VII) Mn(II)	Fe(III) Fe(II)
Electrode potential/V (higher ox. state $\rightleftharpoons$ lower)				
Electrode potential/V ($I_2 + 2e^- \rightleftharpoons 2I^-$)				
Electrode potential/V ($S_2O_8^{2-} + 2e^- \rightleftharpoons 2SO_4^{2-}$)				
Does higher ox. state oxidize I^-?	Prediction			
	Practically			
Does lower state reduce $S_2O_8^{2-}$?	Prediction			
	Practically			
Do you expect catalysis?				
Time for blue colour to appear/s				

3. In a test-tube, mix a few drops of potassium iodide solution with a few drops of dilute sulphuric acid. Add a few drops of one of the solutions containing a transition element in a higher oxidation state. Look for evidence of reaction and fill in your table accordingly.

4. In a test-tube, mix a few drops of potassium peroxodisulphate solution with a few drops of a solution containing the same transition element in a lower oxidation state. Look for evidence of reaction and fill in your table accordingly.

5. Repeat steps 3 and 4 for the other two transition elements. From your results predict which of the given solutions are likely to be catalysts.

   To check whether catalysis actually occurs, proceed as follows. (If you do not have time to check every prediction, share your results with other students.)

6. Fill 4 burettes with one each of the following solutions:

   potassium iodide,   potassium peroxodisulphate,

   starch,   sodium thiosulphate.

7. From the burettes run into a conical flask these volumes:

    10 cm³ of potassium iodide solution,

    10 cm³ of sodium thiosulphate solution,

    5 cm³ of starch solution

8. Run 20 cm³ of potassium peroxodisulphate solution into a boiling-tube.

9. Quickly pour the contents of the boiling-tube into the flask and start the stopclock. Swirl the flask twice to mix the contents and then stand it on the bench while you look for the appearance of a blue colour.

10. As soon as the solution turns blue, stop the clock and record the time taken.

11. Repeat steps 7 to 10, but in addition to the first three reagents, add 5 drops of a solution containing a transition metal ion.

12. If you have time to repeat the procedure before trying another possible catalyst then, of course, your results should be more reliable.

Questions

1. For one of the three transition elements you used, all of the predictions were confirmed by experiment. Which one was this? Describe briefly a possible mechanism (without equations) for catalysis involving this element.

2. One of the three transition elements did not catalyse the reaction. Which one was this? Why could it not function as a catalyst?

3. One of the three transition elements worked well as a catalyst, even though its lower oxidation state did not appear to react with peroxo-disulphate. Which one was this? Suggest a possible explanation. (Hint: look at Experiment 69 if you have done it.)

4. Two students performed this experiment on different days using the same equipment and solutions. They found that their results did not agree very well. Suggest an explanation.

EXPERIMENT 75
Observation and deduction exercise

Aim

The purpose of this experiment is to give
you further practice in the investigation
of unknown substances.

Introduction

The procedure which follows is taken from an A-level paper. If you
are preparing for a practical examination you should ask your teacher
whether you may refer to textbooks and/or notes during this exercise.

A

Requirements

safety spectacles
4 test-tubes in a.rack
3 boiling-tubes
Bunsen burner and bench mat
spatula
beaker, 100 cm³
test-tube holder
salts C and D
sodium hydroxide solution, 2 M NaOH
sulphuric acid, dilute, 1 M H_2SO_4
2 teat-pipettes
potassium iodide solution, 0.6 M KI
sodium thiosulphate solution, 0.1 M $Na_2S_2O_3$
copper powder, Cu
zinc powder, Zn
flame-test rod or wire
watch-glass
hydrochloric acid, concentrated, HCl — — — — — — — — — — — — — — — —
wash-bottle of distilled water
hydrogen peroxide, solution, 1.7 M H_2O_2 — — — — — — — — — — — — —
wood splints
litmus papers, red and blue
other chemicals are available from your teacher

Procedure

You are provided with two salts C and D. Carry out the following tests and
record your observations and inferences in a larger copy of Table 75. Then
answer the questions which follow the table.

Test	Observations	Inferences
1. Warm three-quarters of your sample of C with 4-5 cm³ of aqueous sodium hydroxide.		
2. Add approximately 10 cm³ of dilute sulphuric acid to the remainder of your sample of C in a boiling-tube and warm the mixture in order to dissolve the solid. Use portions of the solution for the following tests:		
(a) To 1-2 cm³ of the solution of C add an equal volume of aqueous potassium iodide. Then add, dropwise, aqueous sodium thiosulphate until there is no further change.		
(b) To 2 cm³ of the solution add a little copper powder and warm.		
(c) To 3-4 cm³ of the solution add a little zinc powder and allow the mixture to stand. It is suggested that you make observations for about 5 minutes and then allow the mixture to stand for a further 30 minutes, making observations from time to time. During this time you should proceed with other tests.		
3. Carry out a flame test on substance D.		
4. Dissolve some D in the minimum quantity of distilled water and use portions of the solution for the following tests:		
(a) Add a few drops of this solution to a mixture of 1-2 cm³ of aqueous potassium iodide with an equal volume of dilute sulphuric acid. Then add aqueous sodium thiosulphate dropwise.		
(b) To 1 cm³ of the solution of D add aqueous sodium hydroxide. Then add dilute sulphuric acid until in excess.		
(c) To about 2 cm³ of the solution of D add an equal volume of dilute sulphuric acid. Then add hydrogen peroxide solution dropwise until there is no further change.		(No inference required.)
(d) Transfer the solution from (c) to a boiling-tube and add about 10 cm³ of aqueous sodium hydroxide and 1 cm³ of hydrogen peroxide solution. Heat the resulting solution.		

The anion in D contains oxygen and a metal. Give the oxidation states of the metal which are indicated in the reactions in 4(a), 4(b) and 4(d).

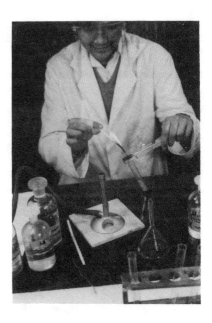

EXPERIMENT 76
Reactions of aluminium

Aim

The purpose of this experiment is to show
how aluminium foil reacts with dilute
acids and alkalis and, when reactions do
occur, to carry out tests on the resulting solutions.

Introduction

In your studies of the periodic table you learned that aluminium oxide is
amphoteric. Other metals with amphoteric oxides such as beryllium and zinc
tend to react with acids and alkalis, so you might also expect aluminium to
react in a similar way.

In addition to observing the reactions of aluminium with dilute acids and
alkalis, you also attempt to interpret some reactions of the resulting
solutions.

Requirements

safety spectacles
6 pieces of aluminium foil, 3 cm x 3 cm
10 test-tubes
test-tube rack
test-tube holder
sodium hydroxide solution, 2 M NaOH — — — — — — — — — — — — — — —
hydrochloric acid, dilute, 2 M HCl
Bunsen burner and bench mat
beaker, 100 cm³
copper(II) chloride solution, 0.1 M $CuCl_2$
mercury(II) chloride solution, 0.1 M $HgCl_2$ — — — — — — — — — — — —
wash-bottle of distilled water
filter funnel
filter paper
sulphuric acid, dilute, 1 M H_2SO_4
sodium carbonate solution, 1 M Na_2CO_3
red and blue litmus papers

Procedure

Reactions of aluminium foil

1. Tear one of the sheets of aluminium foil into smaller pieces and place
    them in a test-tube containing 3-4 cm³ of dilute sodium hydroxide
    solution. Identify any gas given off and record your observations in a
    copy of Results Table 76a. Keep the resulting solution for further tests.

2. Repeat step 1 using dilute hydrochloric acid instead of sodium hydroxide
    solution. If no reaction occurs, heat the mixture gently.

3. Place three pieces of aluminium in a small beaker and just cover them with copper(II) chloride solution. On a fourth piece of aluminium, place three separate drops of mercury(II) chloride solution.

4. After about two minutes, pour away the copper(II) chloride solution and rinse the foil with distilled water. Leave one piece exposed to the air. Put the others in two test-tubes and add sodium hydroxide and dilute hydrochloric acid respectively. Compare what happens with the results of steps 1 and 2.

5. Rinse away the drops of mercury(II) chloride solution with distilled water and leave the foil exposed to the air. Examine it after a few minutes and compare it with the foil treated with copper(II) chloride.

Reactions of the resulting solutions

6. Filter the resulting solutions from steps 1 and 2 (if necessary) and divide each solution into three portions.

7. To separate portions of the solutions, add the following reagents, a little at a time, until they are present in excess:

   (a) dilute sulphuric acid,

   (b) dilute sodium hydroxide solution,

   (c) sodium carbonate solution.

8. Record your observations in a copy of Results Table 76b. If no reaction occurs write NONE in the appropriate box.

Results Table 76a. Reactions of aluminium foil

Reagent	Observations	Identity of any gas given off
Sodium hydroxide solution		
Sodium hydroxide solution after immersion in $CuCl_2$(aq)		
Dilute hydrochloric acid		
Dilute hydrochloric acid after immersion in $CuCl_2$(aq)		
Air after immersion in $CuCl_2$(aq)		
Air after immersion in $HgCl_2$(aq)		

Results Table 76b. Reactions of the aluminium solutions

	Observations using solutions of:	
Reagent	Al in NaOH	Al in HCl
Dilute sulphuric acid		
Sodium hydroxide solution		
Sodium carbonate solution		

You may need to refer to your textbook(s) in order to answer the following questions.

Questions

1. Untreated aluminium has a wide variety of uses which depend, in part, on its resistance to corrosion in normal conditions. However, standard electrode potentials suggest that aluminium is more reactive than iron, which corrodes badly. Explain briefly.

2. How does treatment with a solution of copper(II) chloride or mercury(II) chloride reveal the true reactivity of aluminium?

3. Suggest a reason for the fact that mercury(II) chloride is more effective than copper(II) chloride in promoting a reaction between aluminium and air.

Fig. 42. **Some useful products made from aluminium**

4. Why should you not use washing-soda or some special oven-cleaners on aluminium kitchenware?

5. Write equations for the reactions of aluminium with dilute acids and alkalis.

6. Explain, as far as possible, the observations you have made in Results Table 76b and name the precipitates formed. (Hint: aluminium carbonate is not known.)

EXPERIMENT 77
Anodizing aluminium

Aim

The purpose of this experiment is to demonstrate
the effect of anodizing aluminium on its
electrical conductivity and behaviour towards
dyes.

Introduction

Aluminium is anodized by making it the anode during the electrolysis of a
solution which normally releases oxygen - you will use dilute sulphuric acid.
Instead of oxygen, aluminium ions are released and these are immediately
hydrolysed to give a layer of hydrated oxide. After electrolysis you compare
anodized and unanodized aluminium by testing their conductivities and immer-
sing them in a dye solution.

Requirements

safety spectacles
two pieces of sheet aluminium, 7 cm x 3 cm, mounted on wooden bars
2 beakers, 100 cm^3
forceps or tweezers
cotton-wool
1,1,1-trichloroethane, CH_3CCl_3 — — — — — — — — — — — — — — — — — — —
sulphuric acid, dilute, 1 M H_2SO_4
Bunsen burner, tripod, gauze and bench mat
thermometer, 0-100 °C
DC supply, 12 V
ammeter, 1 A
rheostat, 10 Ω, 4 A
4 connecting leads, two with crocodile clips at one end
2 large pins
dye solution, e.g. alizarin red

Procedure

A. Anodizing

1. Mark one of the wooden supports for the electrodes with a 'plus' sign. This carries the piece of aluminium to be used as the anode.

2. Working at a fume cupboard, remove any grease from the surface of the anode by swabbing with cotton-wool soaked in 1,1,1-trichloroethane and held in forceps or tweezers. From now on, hold the anode only by its wooden support.

3. Pour about 75 cm³ of dilute sulphuric acid into a small beaker, heat to about 40 °C and stand it on the bench.

4. Place the electrodes in the beaker and set up the circuit shown in Fig. 43. Make sure the clean anode is connected to the positive terminal of the supply.

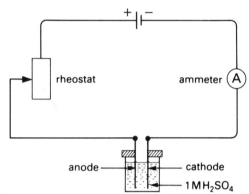

Fig. 43.

5. Switch on the DC supply and adjust the rheostat to give a current of 0.3 A. Maintain this current for at least 15 minutes (longer if it is convenient), adjusting the rheostat if necessary to keep the current constant.

6. Switch off the supply. Disconnect the electrodes, remove them from the beaker and wash them with distilled water. Has their appearance changed?

B. Conductivity testing

7. Clip a large pin to each of the leads you had previously connected to the electrodes. Adjust the rheostat so that when the two pins touch each other, the current is no greater than 1 A.

8. Touch each pin <u>lightly</u> on the surface of the cathode and note the ammeter reading. Repeat for the anode. Is there any difference? Now press the pins more firmly on to the anode surface. If this does not change the ammeter reading, try scratching the surface with the pins.

C. Dyeing

10. Place spots of dye solution on both anode and cathode and leave for 2-3 minutes. (If you have enough dye, you can immerse the electrodes.) Rinse the electrodes with water and inspect them.

Questions

1. How does anodizing affect aluminium with regard to its

    (a) appearance,   (b) conductivity,   (c) susceptibility to dyes?

2. You probably noticed that you had to reduce the variable resistance during the electrolysis in order to prevent the current from falling. Suggest two reasons for this.

3. In what applications might anodizing be

    (a) an advantage,   (b) a disadvantage?

EXPERIMENT 78
Reactions of the oxoacids of
nitrogen and their salts

Aim and introduction

The purpose of this experiment is to illustrate
some of the redox reactions of nitric acid,
nitrates, nitrous acid and nitrites by means
of simple small-scale tests.

Requirements

safety spectacles
protective gloves
6 test-tubes
copper turnings, Cu
nitric acid, concentrated, 16 M HNO_3 — — — — — — — — — — — — —
wash-bottle of distilled water
Bunsen burner and bench protection mat
magnesium ribbon, Mg
nitric acid, dilute, 2 M HNO_3
wood splints
2 boiling-tubes
sodium hydroxide solution, 2 M NaOH — — — — — — — — — — — — — —
Devarda's alloy (Cu 50%, Al 45%, Zn 5%)
aluminium foil, Al
sulphuric acid, dilute, 1 M H_2SO_4
beaker, 250 cm³
ice
spatula
sodium nitrite, $NaNO_2$
potassium iodide solution, 0.5 M KI
potassium manganate(VII) solution, 0.2 M $KMnO_4$
iron(II) sulphate, $FeSO_4 \cdot 7H_2O$

Hazard warning

Concentrated nitric acid and aqueous sodium hydroxide
are corrosive.

Nitrogen dioxide is a toxic gas which may be produced
in reactions of nitrogen compounds.

Therefore you MUST:

WEAR SAFETY SPECTACLES AND PROTECTIVE GLOVES
WORK AT A FUME CUPBOARD

Procedure

A. Reactions of nitric acid

1. Place a single copper turning in a test-tube and carefully add a few drops of concentrated nitric acid. Record your observations in a larger copy of Results Table 78a and identify the gas evolved.

2. Pour about 2 cm³ of concentrated nitric acid into a test-tube and carefully dilute it with an equal volume of distilled water. Drop in 2-3 copper turnings. Identify the gas evolved by observing its colour over the whole length of the test-tube. You may need to warm the tube gently to speed up the reaction.

3. Add about 4 cm³ of 2 M nitric acid to a 4 cm piece of magnesium ribbon in a test-tube. Cork the tube loosely and attempt to identify the gas or gases evolved.

4. Add about 3 cm³ of distilled water to a 4 cm piece of magnesium ribbon in a test-tube and add 1 cm³ of 2 M nitric acid. Cork the tube loosely and attempt to identify the gas evolved.

5. In a boiling-tube, mix 1 cm³ of 2 M nitric acid and 2 cm³ of 2 M sodium hydroxide solution. Add a spatula measure of Devarda's alloy and warm gently. Identify the gas given off.

6. Repeat step 5 using a small piece of aluminium foil instead of Devarda's alloy. Complete your copy of Results Table 78b.

B. Reactions of nitrous acid

7. Prepare a solution of nitrous acid, as follows. Pour about 15 cm³ of dilute sulphuric acid into a boiling-tube and stand it in an ice-bath for about 5 minutes. Dissolve about 1.5 g of sodium nitrite in the minimum quantity of distilled water and cool in the ice-bath. Mix the two cooled solutions.

8. Transfer about 2 cm³ of the nitrous acid solution to a test-tube and warm gently. Record your observations and try to identify the gas evolved.

9. In another test-tube add a few drops of potassium iodide solution to approximately 2 cm³ of nitrous acid solution.

10. Repeat step 9 using a few drops of potassium manganate(VII) solution instead of potassium iodide.

11. Dissolve a few small crystals of iron(II) sulphate in about 2 cm³ of cold distilled water.

   (a) To half of this solution add dilute aqueous sodium hydroxide.

   (b) Add the other half of the iron(II) sulphate solution to an equal volume of nitrous acid solution. Heat the mixture till the colour lightens and then cool. Add sodium hydroxide solution and compare the result with what happened in (a).

12. To a little solid sodium nitrite add 1-2 cm³ of aqueous sodium hydroxide and a small piece of aluminium foil. Heat the mixture and test any gases evolved. Complete your larger copy of Results Table 78b.

Results Table 78a. Reactions of nitric acid and nitrates

Reactants	Observations	Identity of gas
1. Copper and 16 M HNO_3		
2. Copper and 8 M HNO_3		
3. Magnesium and 2 M HNO_3		
4. Magnesium and 0.5 M HNO_3		
5. Devarda's alloy and a nitrate in alkali		
6. Aluminium and a nitrate in alkali		

Results Table 78b. Reactions of nitrous acid and nitrites

Reactants in solution	Observations	Identity of coloured product	Oxidation or reduction of NO_2^-
7. Sodium nitrite and sulphuric acid			
8. Warm nitrous acid			
9. Potassium iodide and nitrous acid			
10. Potassium manganate(VII) and nitrous acid			
11a. Iron(II) sulphate and sodium hydroxide			
11b. Iron(II) sulphate, nitrous acid and sodium hydroxide			
12. Aluminium and a nitrite in alkali			

Questions

To help you with your answers you may need to refer to your textbooks.

1. Write equations for the reactions between copper and nitric acid of different concentrations. Explain your observations of these reactions.

2. Describe and explain the difference between the reactions of magnesium with dilute and very dilute nitric acid.

3. Aluminium is the most powerful reducing agent of the three metals in Devarda's alloy. Suggest reasons why the alloy is more effective than aluminium.

4. Write an equation for the thermal decomposition of nitrous acid. What type of redox reaction is this? Why does nitric acid not undergo this type of reaction?

5. Which of the reactions may be used to distinguish (a) nitrates from nitrites, and (b) nitrates and nitrites from most other compounds.

Investigating some reactions
of the oxo-salts of sulphur

Aim

The purpose of this experiment is to illustrate
the redox reactions of some oxo-anions of
sulphur, some of which can be used as tests for
these ions.

Introduction

You are asked to carry out some simple test-tube reactions on solutions and
solid samples of the following salts:

sodium sulphite, Na_2SO_3,

sodium sulphate, Na_2SO_4,

sodium thiosulphate, $Na_2S_2O_3$,

sodium peroxodisulphate, $Na_2S_2O_8$ (sodium persulphate).

In some cases you may not be able to interpret the reactions fully, but be
sure to record your observations clearly and precisely.

Requirements

safety spectacles
protective gloves
10 test-tubes in rack
sodium sulphite solution, 0.2 M Na_2SO_3
sodium sulphate solution, 0.2 M Na_2SO_4
sodium thiosulphate solution, 0.2 M $Na_2S_2O_3$
sodium peroxodisulphate solution, 0.2 M $Na_2S_2O_8$
     (persulphate)
hydrochloric acid, dilute, 2 M HCl
Bunsen burner and bench protection mat
wood splints
potassium dichromate(VI) solution, acidified, 0.1 M $K_2Cr_2O_7$
strips of filter paper
silver nitrate solution, 0.1 M $AgNO_3$ — — — — — — — — — — — — — — — —
'silver residues' bottles
iodine solution, 0.2 M I_2 in KI(aq)
potassium iodide solution, 0.5 M KI
iron(III) chloride solution, 0.5 M $FeCl_3$
sodium hydroxide solution, 2 M NaOH — — — — — — — — — — — — — — —
test-tube holder and spatula
wash-bottle of distilled water
sodium sulphite-7-water, $Na_2SO_3 \cdot 7H_2O$
sodium sulphate-10-water, $Na_2SO_4 \cdot 10H_2O$
sodium thiosulphate-5-water, $Na_2S_2O_3 \cdot 5H_2O$
sodium peroxodisulphate, $Na_2S_2O_8$ — — — — — — — — — — — — — — —
boiling-tube
sulphur, powdered, S
filter funnel
filter papers

Procedure

Perform tests 1 to 5 on separate portions (about 1 cm³) of <u>solutions</u> of the four oxo-salts provided.

1. Working at a fume cupboard, add an equal volume of dilute hydro-chloric acid, a few drops at a time. Warm <u>gently</u> (do not boil) and test any gas which you can see or smell. Record your observations in a larger copy of Results Table 79 and infer what you can from them.

2. Add about 2 cm³ of silver nitrate solution, a few drops at a time. Pour residues into the bottle provided.

3. Add a few drops of iodine solution.

4. Add a few drops of potassium iodide solution.

5. Add two drops of iron(III) chloride solution, followed by a few drops of dilute hydrochloric acid. Warm gently (do not boil) for half a minute and then add sodium hydroxide solution.

6. Heat small separate portions (about 0.5 g) of the solid oxo-salts in a series of clean test-tubes. Test any gases evolved. Record your observations and inferences in your copy of Results Table 79.

7. Place about 0.5 g of finely powdered sulphur in a boiling-tube and add about 10 cm³ of sodium sulphite solution. Boil the mixture for 2-3 minutes and then filter.

8. Perform tests 1 to 5 on portions of the filtrate from step 7. Record your observations in the last column of Results Table 79 and attempt to identify the new species, X, in the filtrate.

Results Table 79

Test	Observations and inferences				
	Na_2SO_3	Na_2SO_4	$Na_2S_2O_3$	$Na_2S_2O_8$	'X'
1. Dilute hydrochloric acid. Warm.	(a)	(b)	(c)	(d)	
2. Silver nitrate solution.	(e)	(f)	(g)	(h)	
3. Iodine solution in aqueous potassium iodide.	(i)	(j)	(k)	(l)	
4. Potassium iodide solution.	(m)	(n)	(o)	(p)	
5. Iron(III) chloride solution. Acidify, warm, add alkali.	(q)	(r)	(s)	(t)	
6. Effect of heat on solid.	(u)	(v)	(w)	(x)	

Questions

1. With the aid of your textbook(s) interpret your observations as far as you can, writing equations where possible, especially for (a), (c), (i), (k), (l) and (q). Insert oxidation numbers for sulphur in the equations.

2. Which of the reactions illustrate(s) disproportionation?

3. Which of the oxo-salts is: (i) the strongest oxidizing agent, (ii) the strongest reducing agent, and (iii) the most stable salt?

4. Apart from the reactions used in the experiment describe, from your previous knowledge, how you would distinguish between sodium sulphite and sodium sulphate.

5. What is the identity of the new species, X, produced by heating sodium sulphite solution with sulphur?

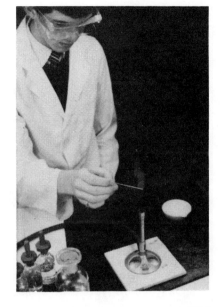

EXPERIMENT 80
Observation and deduction exercise

Aim

The purpose of this experiment is to give
you further practice in the investigation
of unknown substances.

Introduction

The procedure which follows is taken from an A-level paper. If you
are preparing for a practical examination you should ask your teacher
whether you may refer to textbooks and/or notes during this exercise.

A

Requirements

safety spectacles
4 test-tubes in a rack
aqueous solutions of salts B, C and D
2 teat-pipettes
hydrochloric acid, dilute, 2 M HCl
barium chloride solution, 0.5 M $BaCl_2$ — — — — — — — — — — — — — —
lead(II) ethanoate (acetate) solution, 0.25 M $(CH_3CO_2)_2Pb$ — — — — — —
ammonium ethanoate (acetate) solution, 3 M $CH_3CO_2NH_4$
boiling-tube
Bunsen burner and bench-mat
tripod and gauze
litmus papers, blue and red
strips of filter paper
wash-bottle of distilled water
potassium dichromate(VI) solution, 0.5 M $K_2Cr_2O_7$ — — — — — — — — — —
beaker, 100 cm^3
copper(II) sulphate solution, 0.5 M $CuSO_4$
sulphuric acid, dilute, 1 M H_2SO_4
potassium iodide solution, 0.5 M KI
sodium chloride solution, 0.1 M NaCl
silver nitrate solution, 0.1 M $AgNO_3$ — — — — — — — — — — — — —
iron(II) sulphate
potassium thiocyanate solution, 0.5 M KSCN — — — — — — — — — — — —
evaporating dish
hydrochloric acid, concentrated, HCl — — — — — — — — — — — —
watch glass
flame-test wire or rod
other chemicals are available from your teacher

Procedure

You are provided with aqueous solutions, labelled B, C and D, of three salts.
They are compounds of the same three elements.

Carry out the following experiments. Record your observations and inferences
in a copy of Results Table 80, commenting on the types of chemical reactions
involved.

Results Table 80

Test	Observations	Inferences
1. To 2 or 3 cm³ of solution B add a few drops of dilute hydrochloric acid. Now add a few drops of aqueous barium chloride.		
2. To 2 or 3 cm³ of solution B add a few drops of aqueous lead ethanoate (lead acetate). Now add an excess of aqueous ammonium ethanoate (ammonium acetate) and shake the mixture.		
3. To about 5 cm³ of solution C in a boiling-tube add about 10 cm³ of dilute hydrochloric acid. Allow the mixture to stand for a minute or two. Cautiously smell the mixture. Warm, if necessary, and test the gas evolved. Describe below how you performed this test. Method:		
4. (a) In a small beaker place 2 or 3 cm³ of aqueous copper(II) sulphate. Acidify with two drops of dilute sulphuric acid. Now add about 5 cm of aqueous potassium iodide. (b) To the mixture obtained in 4(a) add solution C until in excess, swirling well.		
5. (a) In a small beaker mix about 2 cm³ of aqueous sodium chloride with an equal volume of aqueous silver nitrate. (b) To the mixture obtained in 5(a) add solution C until in excess, swirling well.		
6. Dissolve about 1 g of iron(II) sulphate crystals in dilute sulphuric acid and add a few drops of aqueous potassium thiocyanate. Now add a few drops of solution D.		
7. (a) In a small beaker place about 5 cm³ of aqueous potassium iodide and a few drops of dilute sulphuric acid. Now add about 5 cm³ of solution D. Warm this mixture. (b) To the mixture obtained in 7(a) add solution C until in excess, swirling well.		
8. Evaporate a small quantity of solution B to dryness and perform a flame test on the residue. Describe below how you do this. Method:		

EXPERIMENT 81
Chemical properties of alkanes

Aim

The purpose of this experiment is to test
the reactivity of the alkanes using
cyclohexane as an example.

Introduction

We have chosen cyclohexane as an example of an alkane because it is a liquid,
which makes it easy to handle, and because it is cheap. It has virtually
the same reactions as hexane and is very similar to other alkanes. You use
cyclohexane in five simple test-tube reactions.

Requirements

safety spectacles
access to fume cupboard
protective gloves
cyclohexane (with teat-pipette) ———————————————————————————
hard glass watch-glass
Bunsen burner and bench protection mat
wood splints
1 dry test-tube covered in aluminium foil
5 dry test-tubes with corks to fit
test-tube rack
bromine dissolved in 1,1,1-trichloroethane (with teat-pipette) ─┘
lamp with 100 watt bulb
concentrated ammonia solution ——————————————————————————
bromine water (with teat-pipette) —————————————————————————
dilute sulphuric acid
potassium manganate(VII) solution 0.01 M $KMnO_4$
concentrated sulphuric acid ——————————————————————————————

Hazard warning

Bromine is dangerously toxic and corrosive, especially in
its liquid state. Solutions, such as those used in this
experiment, must also be treated with care. Therefore
you MUST:

DO THE EXPERIMENT IN A FUME CUPBOARD.
KEEP THE TOP ON THE BOTTLE AS MUCH AS POSSIBLE.
WEAR GLOVES AND SAFETY SPECTACLES.

Cyclohexane is very flammable. Therefore you MUST:

KEEP THE TOP ON THE BOTTLE AS MUCH AS POSSIBLE.
KEEP THE BOTTLE AWAY FROM FLAMES.
WEAR SAFETY SPECTACLES.

Concentrated sulphuric acid is very corrosive and reacts violently with water. Therefore you MUST:

WEAR GLOVES AND SAFETY SPECTACLES.
DISPOSE OF UNWANTED ACID BY COOLING AND POURING SLOWLY
INTO AN EXCESS OF WATER.

Procedure

A. Combustion

1. Place your watch glass on a bench protection sheet in the fume cupboard. Put on safety spectacles and make sure the extractor in the fume cupboard is switched on.

2. Using a teat-pipette, place 3-4 drops of cyclohexane on the watch-glass.

3. Stopper and remove the bottle of cyclohexane to a safe place away from the watch-glass and any Bunsen flames.

4. Pull down the front of the fume cupboard leaving a 30 cm opening.

5. Light a long splint and use this to light the cyclohexane. Lower the front of the fume cupboard to a 10 cm opening.

6. Write down, in a larger copy of Results Table 81:

   (a) the colour of the flame,

   (b) whether you can see any soot produced.

B. Reaction of bromine (dissolved in 1,1,1-trichloroethane)

1. Place the test-tube covered with aluminium foil in a rack in the fume cupboard. Put an uncovered tube alongside. Put on safety spectacles and gloves.

2. Using a teat pipette, place approximately 2 cm³ of cyclohexane in each test-tube.

3. Stopper the cyclohexane and remove it to a safe place away from flames.

4. Pull down the front of the fume cupboard leaving a 30 cm opening.

5. Using a teat pipette, place in each tube five drops of a solution of bromine in 1,1,1-trichloroethane.

6. Stopper the bromine bottle.

7. Shine the lamp on both test-tubes for about 3 minutes.

8. A gas is given off during this experiment. Think what gas could be given off and work out a test for the gas. Note the test and its result in your Results Table.

9. Note the appearance of the contents of the clear test-tube.

10. Pour the contents of the test-tube covered with aluminium foil into a clean test-tube. Note its appearance.

C. Reaction of bromine water

1. Place a clean test-tube in a rack in the fume cupboard. Put on safety spectacles and gloves.

2. Using a teat-pipette, place approximately 1 cm³ of cyclohexane in the test-tube.

3. Stopper the bottle of cylcohexane and remove it to a safe place away from the flames.

4. Pull down the front of the fume cupboard leaving a 30 cm opening.

5. Using a teat-pipette, place 5 drops of bromine water in the test-tube.

6. Stopper the bottle of bromine water.

7. Cork and shake the test-tube.

8. Note the appearance of the reaction mixture.

D. Reaction of acidified potassium manganate(VII)

1. Place a test-tube in a rack in the fume cupboard.

2. Using a teat-pipette, place 3-4 drops of cyclohexane in the test-tube.

3. Stopper and remove the bottle of cyclohexane to a safe place, away from flames.

4. Pour into the test-tube approximately 1 cm^3 of dilute sulphuric acid. Shake the mixture.

5. Pour into the test-tube 5-6 drops of potassium manganate(VII) solution and shake the mixture.

6. Note the appearance of the reaction mixture.

E. Reaction of concentrated sulphuric acid

1. Place a test-tube in a rack in the fume cupboard.

2. Pour into the test-tube approximately 1 cm^3 of concentrated sulphuric acid.

3. Pour into the test-tube approximately 1 cm^3 of cyclohexane.

4. Stopper and remove the bottle of cyclohexane to a safe place, away from flames.

5. Note whether the substances mix or form two separate layers.

Results Table 81. Reactions of alkanes

Reaction	Observations

A. Combustion
   Appearance of flame
   Sootiness	

B. Action of bromine (in
   1,1,1-trichloroethane
   1. In dark.
   2. In light.
   Identification of gas	

C. Action of bromine water	
D. Action of acidified potassium manganate(VII)	
E. Action of concentrated sulphuric acid	

Questions

1. Do you consider alkanes to be unreactive? Explain your answer.

2. (a) What are the products of complete combustion of the alkanes?

   (b) Use your data book to write full thermo-chemical equations for the combustion of methane, ethane and propane.

   (c) How do your answers to part (b) relate to the uses of the alkanes?

3. (a) What is meant by the term 'substitution reaction'? Refer to your organic text-book(s).

   (b) Complete the following equation for the substitution reaction between equal amounts of cyclohexane and bromine.

   (c) What conditions favour the reaction shown in (b) above?

   (d) In the presence of excess bromine, it is possible to substitute more than one hydrogen atom per molecule of alkane. Write four equations showing all the possible substitution products of reaction between bromine and methane, CH_4, in sunlight. Name the products.

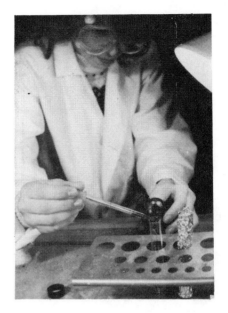

Aim

The purpose of this experiment is to test
the reactivity of the alkenes by carrying
out some test-tube reactions on cyclohexene.

Introduction

You will be using cyclohexene (which is a cycloalkene), because it is one of
the cheapest liquid alkenes. It has virtually the same reactions as hexene
and is similar to other alkenes. You will repeat the same reactions on
cyclohexene that you performed on cyclohexane. This will enable you to
compare the reactivities of the two types of hydrocarbon.

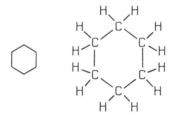

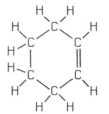

    cyclohexane   cyclohexene

Requirements

Procedure

Results

Work through the procedure in Experiment 81
again except that you substitute cyclohexene
for cyclohexane. Record your results in
another copy of the table under the
heading 'Results Table 82'.

Hazard warning

In Part E, <u>do not shake</u> the mixture of
cyclohexene and sulphuric acid.

Questions

1. Are the alkenes more reactive than the alkanes? Suggest a reason for
    your answer.

2. Why do you think alkenes produce a sootier flame than alkanes?

3. Which test(s) could be used to distinguish between alkanes and alkenes?

EXPERIMENT 83

Hydrolysing organic halogen compounds

Aim

The purpose of this experiment is to find out how the rate of hydrolysis of an organic halogen compound depends on:

(a) the identity of the halogen atom,

(b) the nature of the carbon-hydrogen 'skeleton'.

Introduction

In this experiment, you compare the rates of hydrolysis of 1-chlorobutane, 1-bromobutane, 1-iodobutane and chlorobenzene. A general equation for the hydrolysis is:

$$R-X + H_2O \rightarrow R-OH + H^+ + X^-$$

(where R = alkyl or aryl group; X = halogen atom).

You can follow the rate of the reaction by carrying it out in the presence of silver ions, so that any halide ions produced form a silver halide precipitate.

$$Ag^+(aq) + X^-(aq) \rightarrow AgX(s)$$

Since halogenoalkanes and halogenoarenes are insoluble in water, ethanol is added to act as a common solvent for the halogeno-compounds and silver ions.

Requirements

safety spectacles
Bunsen burner, tripod, gauze and bench protection sheet
  (or a thermostatically controlled water-bath, set at 60 °C)
beaker, 250 cm³
thermometer, 0-100 °C
5 test-tubes fitted with corks
test-tube rack
waterproof marker or chinagraph pencil
protective plastic gloves
measuring cylinder, 10 cm³
ethanol, C_2H_5OH –
1-chlorobutane, C_4H_9Cl ⎫
1-bromobutane, C_4H_9Br ⎬ –
1-iodobutane, C_4H_9I ⎪
chlorobenzene, C_6H_5Cl ⎭
silver nitrate solution, 0.05 M $AgNO_3$
stop-clock (or clock with seconds hand)

202

```
Hazard warning

All organic halogen compounds have harmful vapours and can be
toxic by absorption through the skin.  Some are flammable.
Therefore you MUST:

KEEP THE STOPPERS ON THE BOTTLES AS MUCH AS POSSIBLE

KEEP THE BOTTLES AWAY FROM FLAMES

WEAR SAFETY SPECTACLES AND GLOVES
```

Procedure

1. Set up the apparatus shown in
   Fig. 44 (or switch on your
   electric water-bath).

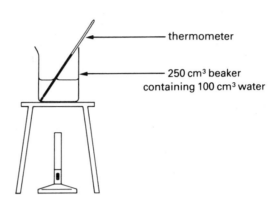

Fig. 44.

2. Pour 2 cm³ of ethanol into each of four test-tubes and mark them with
   the letters A to D.

3. Add 3-4 drops of 1-chlorobutane to A, 3-4 drops of 1-bromobutane to B,
   3-4 drops of 1-iodobutane to C and 3-4 drops of chlorobenzene to D.

4. Pour about 5 cm³ of silver nitrate solution into the fifth test-tube.

5. Stand all the test-tubes in the beaker (or water-bath) and heat to 60 °C.
   Remove the Bunsen burner.

6. Quickly add 1 cm³ of aqueous silver nitrate to each of the tubes A to D
   and start the clock. Shake each tube once to mix the contents, and leave
   in the water with the cork resting loosely on the tube to reduce evaporation.

7. Watch the tubes continuously for about ten minutes and note, in a copy of
   Results Table 83, the time when a cloudy precipitate first appears in each
   tube. If necessary, heat the water to 60 °C again at intervals.

8. Continue observation at intervals for about 30 minutes more, noting any
   further changes in the appearance of the precipitates.

Results Table 83

Reaction	Time for precipi-tate to appear	Observations
A 1-chlorobutane		
B 1-bromobutane		
C 1-iodobutane		
D chlorobenzene		

Questions

1. List the compounds in order of speed of hydrolysis, fastest first.

2. Suggest an explanation for this order in terms of the C-halogen bond
   energies.

3. Write equations for the hydrolysis reactions which occur in this experiment.

EXPERIMENT 84
The Preparing a halogenoalkane

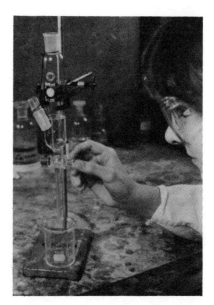

Aim

The purpose of this experiment is to prepare
2-chloro-2-methylpropane and to illustrate
several practical techniques employed in
organic chemistry.

Introduction

In this experiment you prepare 2-chloro-2-methylpropane from 2-methylpropan-
2-ol and hydrochloric acid. The reaction takes place at room temperature
because tertiary alcohols undergo substitution very readily.

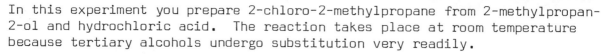

         CH₃   CH₃
          | |
CH₃—C—CH₃ + HCl   →   CH₃—C—CH₃ + H₂O
          | |
         OH   Cl
  2 methylpropan-2-ol   2-chloro-2-methylpropane

In this preparation you will meet the techniques of simple distillation and
solvent extraction, which are commonly used in the purification of an organic
liquid to give the best possible yield.

Requirements

safety spectacles and gloves
measuring cylinder, 25 cm³
2-methylpropan-2-ol, $(CH_3)_3COH$ - — — — — — — — — — — — — — — — — — —
access to balance, sensitivity ± 0.1 g or better
separating funnel, 50 cm³, with stopper
3 retort stands, bosses and clamps
hydrochloric acid, concentrated, HCl — — — — — — — — — — — — — — — —
ground-glass-joint apparatus shown in Fig. 45 (with rubber tubing)
Bunsen burner, tripod and gauze
thermometer, 0-100 °C
conical flask, 100 cm³, with bung
sodium hydrogencarbonate solution, saturated, NaHCO₃
sodium sulphate, anhydrous, Na₂SO₄
spatula
anti-bumping chips

Hazard warning

2-chloro-2-methylpropane and 2-methylpropan-2-ol are flammable.
Concentrated hydrochloric acid is very corrosive.
Therefore you MUST:

KEEP STOPPERS ON BOTTLES AS MUCH AS POSSIBLE

KEEP FLAMMABLE LIQUIDS AWAY FROM FLAMES

WEAR GLOVES AND SAFETY SPECTACLES

Procedure

1. Pour about 9 cm³ of 2-methylpropan-2-ol into a measuring cylinder,
    weigh it, and note its mass in a copy of Results Table 84.

2. Pour the 2-methylpropan-2-ol into a 50 cm³ separating funnel and again
    weigh and record the mass of the measuring cylinder. Then add 20 cm³ of
    concentrated hydrochloric acid about 3 cm³ at a time. After each addition
    hold the stopper and tap securely in place and invert the funnel a few
    times; then, with the funnel in the upright position, loosen the stopper
    briefly to release any pressure.

3. Leave the separating funnel plus contents in the fume cupboard for about
    twenty minutes and shake it gently at intervals.

4. Meanwhile, set up, in a fume cupboard, a clean distillation apparatus,
    as shown in Fig. 45. Remember that the apparatus is a rigid assembly;
    you must be very careful when you clamp it at more than one point, to
    avoid strain and possible breakage.

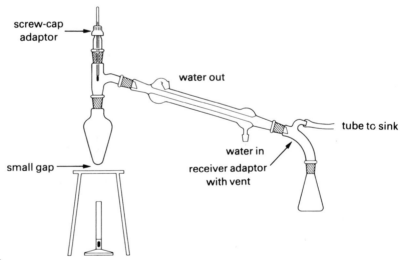

Fig. 45. Distillation

5. Weigh the small flask used as a receiver for the distillate.

6. Allow the layers in the separating funnel to separate; run off and
    discard the lower aqueous layer.

7. Add sodium hydrogencarbonate solution 2 cm³ at a time in order to
    neutralize any excess hydrochloric acid. Shake the funnel carefully
    after each addition and release the pressure of carbon dioxide frequently
    by loosening the stopper. Continue until no more carbon dioxide is
    produced.

8. Allow the layers to separate; run off and discard the lower aqueous
    layer.

9. Run the organic layer into a small, dry conical flask and add about three
    spatula-measures of anhydrous sodium sulphate to dry the organic liquid.
    Cork and swirl the flask occasionally for about five minutes.

10. Carefully pour off (decant) the dried organic liquid from the solid
    sodium sulphate into the pear-shaped flask set up for distillation, as in
    Fig. 45. If you decant slowly, you should be able to separate the solid
    and liquid completely - no solid must enter the distilling flask.

11. Add a few anti-bumping granules to the pear-shaped flask and
    distil the 2-chloro-2-methylpropane, collecting the fraction in
    the range 47 to 53 °C into the pre-weighed conical flask. Heat
    gently at first and then more strongly but only just strongly
    enough to keep the product distilling at about 1-2 drops per second.

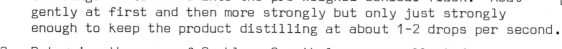

12. Determine the mass of 2-chloro-2-methylpropane collected.

Results Table 84

Mass of measuring cylinder + 2-methylpropan-2-ol	g
Mass of measuring cylinder after emptying	g
Mass of 2-methylpropan-2-ol	g
Mass of collecting flask	g
Mass of collecting flask + 2-chloro-2-methylpropane	g
Mass of 2-chloro-2-methylpropane	g

Questions

1. From the chemical equation, calculate the maximum mass of 2-chloro-2-
    methylpropane that could be formed from the mass of 2-methylpropan-2-ol
    you used.

2. Calculate the percentage yield of 2-chloro-2-methylpropane using the
    expression:

    $$\% \text{ yield} = \frac{\text{actual mass of product}}{\text{maximum mass of product}} \times 100$$

3. Why is sodium hydrogencarbonate used to remove acid impurities, rather
    than a stronger alkali such as sodium hydroxide?

EXPERIMENT 85
Chemical properties of ethanol

Aim

The purpose of this experiment is to study some of the reactions of ethanol, a typical primary alcohol.

Introduction

The reactions to be investigated, with approximate times required, are:

A. Solubility in water   -   5 minutes

B. Mild oxidation   -   30 minutes

C. Further oxidation   -   40 minutes

D. Triiodomethane reaction   -   5 minutes

E. Reaction with sodium   -   5 minutes

F. Esterification   -   5 minutes

G. Dehydration   -   30 minutes

Your teacher may want you to divide up the experiment among a group, with each member reporting back on two parts. If you are working alone, you could omit B or C and G, but ask your teacher for the specimen results.

You will use Fehling's solution in some of your tests. This solution contains a blue copper(II) compound which, in the presence of a reducing agent, changes to a red copper(I) compound.

$$Cu^{2+} + e^- \rightarrow Cu^+ \rightarrow Cu_2O$$
$$\text{blue} \qquad\qquad\qquad \text{red}$$

You will also use an ammoniacal solution of silver oxide (Tollens' reagent) which, in the presence of a reducing agent, produces a 'silver mirror' (and/or a grey precipitate):

$$Ag^+ + e^- \rightarrow Ag(s)$$

Each of these tests confirms the presence of a reducing agent.

Requirements

A. safety spectacles
   test-tube
   distilled water
   universal indicator
   ethanol, C_2H_5OH — — — — —

B. safety spectacles
   protective gloves
   ground-glass-joint
      apparatus in Fig. 46
   measuring cylinder, 10 cm^3
   sulphuric acid, dilute, 1 M H_2SO_4
   spatula and 3 teat-pipettes
   small funnel, wide stem
   sodium dichromate(VI),
      solid, $Na_2Cr_2O_7$
   balance, ± 0.1 g
   anti-bumping granules
   ethanol, C_2H_5OH — — — — —
   2 retort stands,
      bosses and clamps
   Bunsen burner, tripod,
      gauze, bench mat
   test-tube and holder
   boiling-tube
   Fehling's solution 1
   Fehling's solution 2
   sodium hydroxide
      solution, 2 M NaOH — — —
   silver nitrate solution,
      0.05 M $AgNO_3$
   ammonia solution, 2 M NH_3
   beaker, 250 cm^3

C. safety spectacles
   protective gloves
   ground-glass-joint
      apparatus in Fig. 46
   measuring cylinder, 10 cm^3
   sulphuric acid, dilute, 1 M H_2SO_4
   spatula and teat-pipette
   small funnel, wide stem
   sodium dichromate(VI),
      solid, $Na_2Cr_2O_7$ — — — —
   balance, ± 0.1 g
   sulphuric acid,
      concentrated, H_2SO_4 — —
   anti-bumping granules
   ethanol, C_2H_5OH — — — — —
   2 retort stands,
      bosses and clamps
   Bunsen burner, tripod,
      gauze, bench mat
   universal indicator papers
   test-tube
   sodium carbonate,
      anhydrous, Na_2CO_3

D. safety spectacles
   test-tube and 2 teat-pipettes
   ethanol, C_2H_5OH — — — — —
   iodine solution, 10% I_2 in KI
   sodium hydroxide
      solution, 2 M NaOH — — —

E. safety spectacles
   test-tube, in rack
   ethanol, C_2H_5OH — — — — —
   sodium, Na, small
      pieces under oil — — — — —
   forceps
   filter paper
   splint
   watch-glass

F. safety spectacles
   test-tube and holder
   ethanol, C_2H_5OH
   ethanoic acid,
      glacial, CH_3CO_2H — — — —
   sulphuric acid,
      concentrated, H_2SO_4 — — — —
   Bunsen burner, and bench mat
   beaker, 100 cm^3
   sodium carbonate
      solution, 1 M Na_2CO_3
   teat-pipette

G. safety spectacles
   3 test-tubes, with corks
   ceramic wool
   ethanol, C_2H_5OH — — — — —
   teat-pipette
   pumice stone, 4-8 mesh
   2 retort stands
      bosses and clamps
   conical flask, 100 cm^3
   delivery tubes and
      bungs as in Fig. 48
   water trough
   Bunsen burner and
      bench mat
   bromine water, $Br_2(aq)$ — — —
   potassium manganate(VII)
      solution, 0.01 M $KMnO_4$

Hazard warning

Ethanol is very flammable. Therefore you MUST:

KEEP THE STOPPER ON THE BOTTLE AS MUCH AS POSSIBLE.
KEEP THE BOTTLE AWAY FROM FLAMES.
WEAR SAFETY SPECTACLES.

Bromine and glacial ethanoic acid (acetic acid) have dangerous
fumes and burn the skin. Therefore you MUST:

KEEP THE STOPPERS ON THE BOTTLES AS MUCH AS POSSIBLE.
WEAR GLOVES AND SAFETY SPECTACLES.

Concentrated sulphuric acid is very corrosive and reacts violently
with water. Therefore you MUST:

WEAR SAFETY SPECTACLES AND GLOVES.
MOP UP MINOR SPILLAGES WITH PLENTY OF WATER.
DISPOSE OF UNWANTED RESIDUES BY COOLING AND POURING SLOWLY
INTO AN EXCESS OF WATER.

Tollens' reagent for the silver mirror test becomes explosive
if dry. Therefore you MUST:

WASH AWAY THE SOLUTIONS FROM EXPERIMENTS A AND B IMMMEDIATELY
YOU HAVE FINISHED.

Sodium is extremely dangerous. Therefore you MUST:

USE AS LITTLE AS POSSIBLE (1 mm cube).
USE FORCEPS TO HANDLE IT.
KEEP IT UNDER OIL
WEAR SAFETY SPECTACLES.

Sodium dichromate(VI) is a powerful oxidant and can damage
the skin. Therefore you MUST:

WEAR SAFETY SPECTACLES AND GLOVES.

Procedure

A.   Solubility in water

1.   Pour about 1 cm³ of distilled water into a test-tube, add a few drops of
     universal indicator and shake gently.

2.   Add about 1 cm³ of ethanol and shake the mixture. Note, in a copy of
     Results Table 85, whether the addition of ethanol has any effect on the
     colour of universal indicator. (Your distilled water may be weakly
     acidic due to the absorption of atmospheric carbon dioxide.)

B.   Mild oxidation

1.   Into a pear-shaped flask, pour 10 cm³ of 1 M sulphuric acid. Using a
     small wide-stemmed funnel, add 3.0 g of sodium dichromate(VI) and
     2-3 anti-bumping granules.

3.   Swirl the flask gently until all the sodium dichromate(VI) has dissolved.

4.   Slowly add 5 cm³ of ethanol and swirl to mix.

5.   Set up the apparatus shown in Fig. 46 (next page). Ensure that water
     enters the condenser from the bottom and leaves at the top.

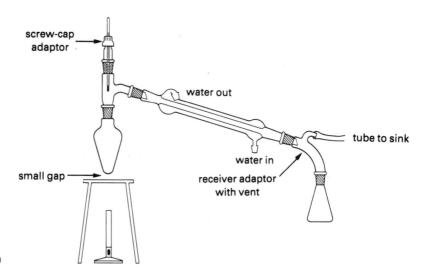

Fig. 46.   Distillation

6.   Heat very gently until 2-3 cm³ of liquid has distilled over.

7.   Keep the distillate and test it in the following ways:

    (a)   Smell cautiously (compare with ethanol).

    (b)   Transfer about 1 cm³ of the distillate to a test-tube. Add about
         1 cm³ of Fehling's solution 1 followed by 1 cm³ of Fehling's
         solution 2. Boil gently. Note your observations in a copy of
         Results Table 85.

    (c)   Pour about 5 cm³ of silver nitrate solution into a boiling-tube.
         Add one drop of sodium hydroxide solution. Drop by drop, add
         aqueous ammonia until the precipitate disappears. Add 2-3 drops of
         the distillate and warm the tube in a beaker containing hot water.
         Note your observations. (This is called the 'silver mirror test'
         or the 'Tollens test'.)

         (DO NOT KEEP THIS SOLUTION - IT BECOMES EXPLOSIVE ON EVAPORATION.)

C.   Further oxidation

1.   Into a pear-shaped flask, pour 10 cm³ of 1 M sulphuric acid. Through a
    wide-stemmed funnel add 5 g of sodium dichromate(VI) and 2 or 3 anti-
    bumping granules.

2.   Swirl the flask gently until all the sodium
    dichromate(VI) has dissolved.

3.   With care, add 2 cm³ concentrated sulphuric acid.

4.   Cool the flask under a running tap.

5.   Set up the apparatus shown in
    Fig. 47, preferably in a fume
    cupboard.

6.   Drop by drop, add 1 cm³ of ethanol
    down the condenser.

7.   Boil gently under reflux for 20
    minutes. (While you are waiting
    you could do reactions D, E and F.)

8.   Rearrange your equipment so it is
    set up as in Fig. 46 above.

9.   Distil 2-3 cm³ of liquid.

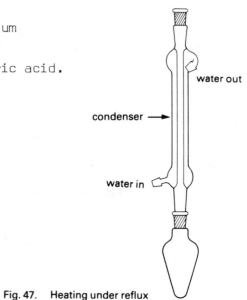

Fig. 47.   Heating under reflux

10. Test the distillate as follows, and note your observations.

    (a) Smell cautiously (compare with ethanol).

    (b) Add a drop to moistened universal indicator paper.

    (c) Add a few drops to about 1 g of solid sodium carbonate.

D. Triiodomethane (iodoform) reaction*

1. In a test-tube, mix 5 drops of ethanol and 1 cm³ of iodine solution.

2. Drop by drop, add sodium hydroxide solution until the brown colour almost disappears. If you observe no other change, warm the tube in a beaker of hot water.

3. Note what happens and smell the product cautiously.

*This reaction is not given by all alcohols; ask your teacher for more detail later. An alternative method, giving the same result, uses potassium iodide and sodium chlorate(I) (hypochlorite).

E. Reaction with sodium (WORK AT A FUME CUPBOARD, WITH YOUR TEACHER PRESENT)

1. Pour about 1 cm³ of ethanol into a test-tube.

2. Using forceps, pick up a 1 mm cube of sodium and remove the oil from its surface on filter paper. Drop the sodium into the ethanol.

3. With the front of the fume cupboard pulled down as far as is practically possible, test the gas with a lighted splint.

4. Pour a little of the product from step 2 on to a watch-glass, leave in the fume cupboard and allow to evaporate. Describe what remains.

F. Esterification

1. Into a test-tube, pour 2 cm³ of ethanol and 1 cm³ of glacial ethanoic acid.

2. WITH CARE, add 2-3 drops of concentrated sulphuric acid.

3. Warm gently for a few minutes but do not boil.

4. Pour the product carefully into a beaker containing sodium carbonate solution. Stir and smell. Note your observations.

G. Dehydration

1. Push enough ceramic wool down to the bottom of the test-tube to fill it to a depth of 2 cm.

2. Using a pipette, drop 2 cm³ of ethanol onto the ceramic wool and allow to soak into the wool.

3. Fill up the rest of the test-tube with the pumice stone.

4. Set up the apparatus shown in Fig. 48, in the fume cupboard, making sure clamp (A) is at the open end of the test-tube.

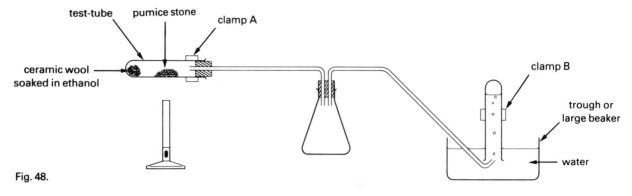

Fig. 48.

5. Holding the Bunsen burner in your hand, heat the pumice stone quite strongly (without melting the tube!) and occasionally heat the ethanol <u>gently</u> to drive the vapour over the hot pumice.

6. Allow the first bubbles to escape (this is displaced air) before collecting the gas over water.

7. Collect a tube of gas (two if possible), cork and place in a rack.

8. Test separate tubes of gas as follows and note your observations.

    (a) Shake with 1-2 drops of bromine water.

    (b) Shake with 1-2 drops of acidified potassium manganate(VII) solution.

Results Table 85. Reactions of ethanol

Property/Reaction	Observations
A. <u>Solubility in water</u> pH of solution	

B. <u>Mild oxidation</u>
   (a) Smell
   (b) Fehling's solution
   (c) Silver mirror test	

C. <u>Further oxidation</u>
   (a) Smell
   (b) Universal indicator paper
   (c) Sodium carbonate	

D. <u>Triiodomethane reaction</u>	
E. <u>Reaction with sodium</u>	
F. <u>Esterification</u>	

G. <u>Dehydration</u>
   (a) Bromine water
   (b) Acidified potassium
       manganate(VII) solution	

Questions

1. Which of the reactions of ethanol produced a reducing agent?
   Use your textbook to find out its name and formula.

2. Why did the orange dichromate(VI) turn green in reactions B and C?

3. Which of the reactions of ethanol produced an acidic compound? The smell
   should give you a clue as to what it might be.

4. What conditions and relative proportions of reactants are used in the
   oxidation of ethanol to favour the production of (a) ethanal,
   (b) ethanoic acid?

5. In the reaction with sodium, what type of fission has taken place in the
   ethanol molecule - is it C—OH or CO—H fission? Explain.

6. What functional group must be in the product of dehydration of ethanol?

7. Which of the reactions produced a pleasant, sweet-smelling compound?

EXPERIMENT 86
Chemical properties of phenol

Aim

The purpose of this experiment is to see the effect of the benzene ring on the behaviour of the hydroxyl group, and vice versa.

Introduction

The reactions you investigate are with the following reagents:

A. Water

B. Sodium hydrogencarbonate

C. Sodium hydroxide

D. Sodium

E. Bromine water

F. Neutral iron(III) chloride solution

Requirements

safety spectacles
protective gloves
6 test-tubes, with corks
test-tube holder and rack
wash-bottle of distilled water
2 beakers, 250 cm^3
phenol, C_6H_5OH —
spatula and wood splint
Bunsen burner, tripod, gauze and bench mat
universal indicator solution
sodium hydrogencarbonate, $NaHCO_3$
sodium hydroxide solution, 2 M NaOH — — — — — — — — — — — — —
hydrochloric acid, concentrated, HCl — — — — — — — — — — — — —
teat-pipette
thermometer, 0-100°C
sodium, Na (1 mm cubes) — — — — — — — — — — — — — — — — — —
forceps
filter paper
bromine water, $Br_2(aq)$ — — — — — — — — — — — — — — — — —
iron(III) chloride solution, 0.5 M $FeCl_3$
sodium carbonate solution, 1 M Na_2CO_3

Hazard warning

Both phenol and bromine water attack the skin
and give off irritating vapours. Therefore you MUST:

WEAR GLOVES AND SAFETY SPECTACLES
AVOID CONTACT WITH SKIN
KEEP STOPPERS ON BOTTLES AS MUCH AS POSSIBLE

Sodium is dangerously reactive. Therefore you MUST:

USE AS LITTLE AS POSSIBLE (1 mm CUBE)
HANDLE WITH FORCEPS
KEEP IT UNDER OIL
WEAR SAFETY SPECTACLES

Procedure

A. Solubility in water

1. Pour about 5 cm³ of water into a test-tube and add a heaped
    spatula-measure (about 1.5 g) of phenol. Cork and shake the
    tube. Note, in a copy of Results Table 86, whether phenol
    dissolves in water.

2. Add a further 3-4 measures of phenol, shake and note your observations.

3. Place the test-tube in a beaker of hot water for a few minutes, shake
    the tube and note your observations.

4. Allow the solution in the test-tube to cool and note your observations.

5. Add a few drops of universal indicator solution to the phenol-water
    mixture in one test-tube and distilled water in another.

B. Reaction with sodium hydrogencarbonate

1. Into a test-tube pour about 1 cm³ of water and add 2-3 small crystals of
    phenol. Cork and shake until the phenol has dissolved.

2. Add about 0.5 g of solid sodium hydrogencarbonate and note your
    observations.

Results Table 86. Reactions of phenol

Test	Observations

A. Solubility in water
    (a) a little phenol
    (b) a lot of phenol
    (c) pH of solution	

B. Reaction with sodium hydrogencarbonate	
C. Reaction with sodium hydroxide Subsequent addition of hydrochloric acid	
D. Reaction with sodium	
E. Reaction with bromine water	
F. Action of neutral iron(III) chloride solution	

C. Reaction with sodium hydroxide

1. Pour about 5 cm³ of 2 M sodium hydroxide into a test-tube and add a
    spatula-measure of phenol. Cork and shake the test-tube.

2. Add 3 more measures of phenol, cork and shake. Compare the solubility
    of phenol in water from experiment A. Note your observations.

3. Drop by drop, add 2 cm³ of concentrated hydrochloric acid. Shake the
    test-tube and note your observations.

D. Reaction with sodium   (YOUR TEACHER MUST SUPERVISE THIS EXPERIMENT)

1. In a fume cupboard, heat about 100 cm³ water in a beaker to about 60 °C.

2. Put a spatula-measure of phenol in a dry test-tube, and stand it in the hot water until the phenol melts.

3. Using forceps, pick up a 1 mm cube of sodium and remove the excess oil from its surface on a piece of filter paper.

4. Use a holder to remove the test-tube from the hot water and, with the front of the fume cupboard down as far as is practically possible, drop the dry sodium into the molten phenol.

5. Apply a lighted splint to the mouth of the test-tube.

E. Reaction with bromine water

1. Into a test-tube pour about 5 cm³ of water and add a spatula-measure of phenol.

2. Cork and shake until the phenol has dissolved.

3. Pour half the solution into another test-tube for reaction F.

4. Add about six drops of bromine water to the aqueous phenol, shaking the test-tube after the addition of each drop. Note your observations.

F. Reaction with neutral iron(III) chloride solution

1. Into a test-tube, pour about 1 cm³ of iron(III) chloride solution. Add sodium carbonate solution, drop by drop, until a trace of the brown precipitate just remains after shaking.

2. Add a few drops of this 'neutral' iron(III) chloride solution to the phenol solution prepared in Reaction E. Note your observations.

3. Add a few drops of 'neutral' iron(III) chloride solution to about 1 cm³ of ethanol. Note the difference between ethanol and phenol in this test.

Questions

1. What does the smell of phenol remind you of?

2. Which tests distinguish between ethanol and phenol?

3. Explain why phenol is more soluble in sodium hydroxide than in water.

4. Which tests indicate that phenol is a stronger acid than ethanol?

5. Which test indicates that phenol is a weaker acid than dilute mineral acids such as HCl, H_2SO_4 and HNO_3?

6. How do you account for the fact that phenol decolorizes bromine water and forms a white precipitate whereas ethanol does not?

EXPERIMENT 87
Reactions of amines

Aim

The purpose of this experiment is to study some of the chemical properties of butylamine (an alkylamine) and phenylamine (an arylamine) and to compare them with the properties of ammonia.

This will enable you to see how the nature of the carbon-hydrogen skeleton affects the properties of the amine group and how the amine group affects the properties of the benzene ring.

Introduction

You will be carrying out some test-tube reactions on butylamine, phenylamine and aqueous ammonia.

ammonia   butylamine   phenylamine

We have chosen butylamine as a typical alkylamine because it is a liquid and not too volatile. We have chosen phenylamine, which is also a liquid, as a typical arylamine.

The reactions or properties to be investigated are:

A   Solubility in water

B   Reaction with hydrochloric acid

C   Reaction with copper(II) sulphate

D   Reaction with nitrous acid

E   Reaction with bromine water

Since nitrous acid is unstable, it is made *in situ* for reaction D by adding sodium nitrite to dilute hydrochloric acid.

$$NaNO_2(aq) + HCl(aq) \rightarrow NaCl(aq) + HNO_2(aq)$$
$$\text{nitrous acid}$$

Requirements

safety spectacles and protective gloves
6 test-tubes, with bungs or corks and test-tube rack
phenylamine, $C_6H_5NH_2$, with teat-pipette – – – – – – – – – – – –
wash-bottle of distilled water
butylamine, $C_4H_9NH_2$, with teat-pipette – – – – – – – – – – – –
ammonia solution, 2 M NH_3
universal indicator paper
glass stirring rod and spatula
hydrochloric acid, concentrated, HCl, with teat-pipette – – – – –
3 watch-glasses
copper(II) sulphate solution, 1 M $CuSO_4$

Requirements (cont.)

4 boiling-tubes, with bungs or corks
labels for test-tubes and beakers
crushed ice
sodium chloride, NaCl
2 beakers, 600 cm³
spatula
ammonium chloride, NH₄Cl
sodium nitrite, NaNO₂
6 beakers, 100 cm³
phenol, C₆H₅OH —
naphthalen-2-ol, C₁₀H₇OH —
sodium hydroxide solution, 2 M NaOH — — — — — — — — — — — — — — — — —
thermometer, 0-100 °C
limewater and wood splint
Bunsen burner, tripod, gauze and bench mat
bromine water, Br₂(aq)- —

Hazard warning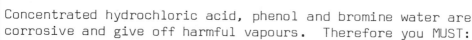

Phenylamine is toxic, by ingestion and by skin absorption.

Phenylamine and butylamine are flammable and give off
harmful vapours.

Concentrated hydrochloric acid, phenol and bromine water are
corrosive and give off harmful vapours. Therefore you MUST:

WEAR SAFETY SPECTACLES AND GLOVES
WORK IN A FUME CUPBOARD WHERE POSSIBLE
KEEP BOTTLES AWAY FROM FLAMES
KEEP STOPPERS ON BOTTLES AS MUCH AS POSSIBLE

Procedure

A. Solubility in water

1. Into a test-tube add 2 drops of phenylamine, 10 drops of water
    and shake. Note whether the amine appears to dissolve.

2. Using a clean glass rod, test one drop of the solution with
    universal indicator paper. Compare the pH of the solution with that
    of distilled water alone and that of ammonia solution. Record your
    observations in a copy of Results Table 87.

3. Repeat steps 1 and 2 using butylamine instead of phenylamine. Also test
    the pH of ammonia solution.

B. Reaction with hydrochloric acid

1. Into a test-tube, add 5 drops of phenylamine, 5 drops of water
    and shake.

2. Drop by drop, add enough concentrated hydrochloric acid to obtain a
    clear solution (about 5 drops).

3. In a fume cupboard, pour the contents of the test-tube on to a watch-
    glass and, if the product is not already solid, allow to evaporate
    slowly. Note your observations.

4. Repeat steps 1 to 3 using butylamine.

5. Repeat steps 2 and 3, starting with 10 drops of ammonia solution.

C. Reaction with copper(II) sulphate solution

1. Into a test-tube, add about 2 cm³ of copper(II) sulphate solution.

2. Drop by drop add phenylamine, shaking the mixture all the time until the phenylamine is present in excess. Note your observations.

3. Repeat with butylamine and aqueous ammonia.

D. Reaction with nitrous acid

You carry out the initial reaction at low temperature (< 5 °C), and then do further tests on the product (if any).

1. Prepare acidic solutions of salts of ammonia, phenylamine and butylamine as follows, and label them clearly.

   (a) In a boiling-tube, dissolve about 5 spatula-measures (~ 5 g) of ammonium chloride in about 15 cm³ of distilled water.

   (b) Into another boiling-tube, pour about 10 cm³ of water and add about 1 cm³ of phenylamine. A little at a time, carefully add about 4 cm³ of concentrated hydrochloric acid, shaking gently to obtain a clear solution.

   (c) Repeat (b) above, using butylamine instead of phenylamine.

2. For purposes of comparison, prepare a fourth labelled tube containing 15 cm³ of water and 2 drops of concentrated hydrochloric acid.

3. Stand the four labelled tubes in a beaker containing a 'freezing mixture' of crushed ice and salt (sodium chloride).

4. Into a fifth boiling-tube, pour about 20 cm³ of water and add two spatula-measures (1.5 g) of sodium nitrite. Cork the tube, shake to dissolve, and stand it in the freezing mixture to cool.

5. While the tubes are cooling, label 6 small beakers P_1, P_2, P_3, N_1, N_2 and N_3. (P for phenol, N for naphthalen-2-ol.)

6. Into beaker P_1, put a spatula-measure of phenol and about 10 cm³ of sodium hydroxide solution. Stir to dissolve the solid, and add about 25 cm³ of water. Divide the solution into 3 portions in beakers P_1, P_2 and P_3.

7. Into beaker N_1, put a spatula-measure of naphthalen-2-ol and about 10 cm³ of sodium hydroxide solution. Stir to dissolve the solid, and add about 25 cm³ of water. Divide the solution into 3 portions in beakers N_2, N_2 and N_3.

8. Check that the temperature of the cooled solutions (prepared in steps 1 to 4) is below 5 °C, then add one-quarter of the sodium nitrite solution to each of the other four tubes.

9. Shake each of the four boiling-tubes and look for signs of a reaction between nitrous acid (formed from the nitrite and excess acid) and the salts of the amines or ammonia. In particular, look for any evolution of gas in excess of that produced by the decomposition of nitrous acid alone in the fourth tube.

10. If a gas is evolved, test it with limewater and with a lighted splint.

11. Into beaker P_1, pour $\frac{1}{3}$ of the phenylamine/nitrous acid mixture.
    Into beaker P_2, pour $\frac{1}{3}$ of the butylamine/nitrous acid mixture.
    Into beaker P_3, pour $\frac{1}{3}$ of the ammonia/nitrous acid mixture.

    Swirl the beakers to mix, and note your observations.

12. Into beaker N_1, pour $\frac{1}{3}$ of the phenylamine/nitrous acid mixture.
    Into beaker N_2, pour $\frac{1}{3}$ of the butylamine/nitrous acid mixture.
    Into beaker N_3, pour $\frac{1}{3}$ of the ammonia/nitrous acid mixture.

    Swirl the beakers to mix, and note your observations.

13. Stand the four labelled boiling-tubes containing the unused nitrous acid and salt mixtures in a beaker of hot water. Look for any evolution of gas in excess of the brown fumes obtained from nitrous acid alone. Test any gas with limewater and a lighted splint.

E. Reaction with bromine water

1. Into a test-tube, place 5 drops of phenylamine and carefully add concentrated hydrochloric acid, dropwise, to obtain a clear solution.

2. Drop by drop, add about 1 cm³ of bromine water. Note your observations.

3. Repeat steps 1 and 2 using butylamine. Check that the solution is acidic; if it is not, add more acid.

4. In a test-tube, dissolve a spatula-measure of ammonium chloride in about 3 cm³ of water. Add bromine water as before and note your observations.

Results Table 87

Test	Phenylamine	Butylamine	Ammonia
A. (a) Solubility in water			
(b) pH of solution			
B. (a) Reaction with hydrochloric acid			
(b) Evaporation of water from product			
C. Reaction with copper(II) sulphate solution			
D. (a) Reaction with cold nitrous acid			
(b) Reaction of product from (a) with (i) phenol (ii) naphthalen-2-ol			
(c) Effect of heat on product from (a)			
E. Reaction with bromine water			

Questions

1. Are butylamine and phenylamine stronger or weaker bases than ammonia?

2. How do you account for the different solubility of the amines in water and in dilute hydrochloric acid?

3. Which reactions might enable you to distinguish between a primary alkylamine and a primary arylamine?

EXPERIMENT 88
Observation and deduction exercise

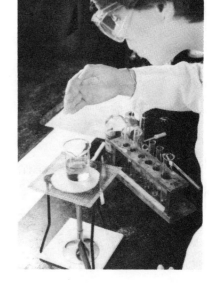

Aim and introduction

This experiment is intended primarily for
students preparing for an A-level practical
examination in which one of the questions
could require the identification of an
unknown compound by means of specified
tests. The procedure which follows is
taken from two A-level practical examination papers; read it carefully and
report fully.

Requirements

safety spectacles
protective gloves
6 test-tubes in a rack
2 wood splints
test-tube holder
substance C –
substance D –
substance E
wash-bottle of distilled water
spatula
sodium carbonate, anhydrous, Na_2CO_3
2 beakers, 100 cm³
2 beakers, 250 cm³
phosphorus pentachloride, PCl_5 – – – – – – – – – – – – – – – – – – –
sulphuric acid, concentrated, H_2SO_4 – – – – – – – – – – – – – – – –
Bunsen burner, tripod, gauze and bench mat
sodium carbonate solution, 1 M Na_2CO_3
potassium dichromate(VI) solution, 0.1 M $K_2Cr_2O_7$ – – – – – – – – – –
sulphuric acid, dilute 1 M H_2SO_4
sodium hydroxide solution, 2 M NaOH – – – – – – – – – – – – – – – –
potassium iodide, KI

sodium chlorate(I) solution (hypochlorite), NaClO – – – – – – – – – –

bromine water, Br_2(aq) –

silver nitrate solution, 0.05 M $AgNO_3$
nitric acid, dilute, 2 M HNO_3
ammonia solution, 2 M NH_3
hydrochloric acid, concentrated, HCl
crushed ice
sodium nitrite, $NaNO_2$
phenol, C_6H_5OH

Procedure

You are provided with three organic compounds, labelled C, D and E. C and D contain the elements carbon, hydrogen and oxygen only. Carry out the following experiments. Record your observations and inferences in (larger copies of) the tables provided. Comment on the types of chemical reaction occurring and, where possible, deduce the functional groups present in these compounds.

Results Table 88a

Test	Observations	Inferences
1.(a) Place 1 cm³ of C in a test-tube and add an equal volume of water. Now add a little anhydrous sodium carbonate. (b) Repeat test 1.(a) using D.		
2. (The reactions 2.(a) and 2.(b) should be performed in a fume cupboard.) (a) Place 2 or 3 cm³ of C in a dry beaker. Add a little phosphorus pentachloride. (CARE) (b) Repeat test 2.(a) using D. (CARE)		
3. Mix, in a small beaker, about 2 cm³ of each of C, D and concentrated sulphuric acid (CARE). Warm gently but do not boil. Pour the mixture into an excess of aqueous sodium carbonate in a large beaker. Smell the product.		
4. Mix about 5 cm³ of D with an equal volume of aqueous potassium dichromate in a test-tube. Pour the mixture into about 10 cm³ of nearly boiling dilute sulphuric acid in a small beaker. Smell the mixture.		
5. In a small beaker dissolve a few crystals of potassium iodide in about 10 cm³ of D. Add a few drops of aqueous sodium hydroxide. Now add 2 or 3 cm³ of aqueous sodium hypochlorite. Warm the mixture but do not boil. Cautiously smell the products.		
6. Make a solution of E in distilled water and use portions for the following tests: To 2-3 cm³ of the solution add aqueous bromine. To 2-3 cm³ of the solution add aqueous silver nitrate. Then add dilute nitric acid. Finally add dilute aqueous ammonia.		
7. Dissolve a little E in about 1 cm³ of concentrated hydrochloric acid and dilute to about 4 cm³ with distilled water. Cool the tube in an ice-water mixture and add a few drops of aqueous sodium nitrite (to be prepared by dissolving sodium nitrite in distilled water). Leave the tube in the ice-water mixture. Dissolve a few crystals of phenol (CARE) in 7-8 cm³ of aqueous sodium hydroxide, cool this solution, and then add it to the cold solution prepared as above.		

1. Complete the following table as far as possible.

Results Table 88b

	REASONING
Functional group in C:	
Functional group in D:	

2. Give the structural features of compound E.

EXPERIMENT 89

Reactions of aldehydes and ketones

Aim

The purpose of this experiment is to compare some reactions of ethanal and propanone.

Introduction

We have chosen ethanal and propanone as relatively safe examples of aldehydes and ketones to illustrate their reactions in simple test-tube experiments.

$$CH_3 \diagdown C=O \diagup H$$
ethanal

$$CH_3 \diagdown C=O \diagup CH_3$$
propanone

The reactions or properties to be investigated are as follows:

A Addition

B Condensation
   (Addition - elimination)

C Reaction with alkali

D Oxidation

E Triiodomethane (iodoform) reaction

You may have used Fehling's solution and Tollen's reagent (ammoniacal silver nitrate) in Experiment 85 to test for an aldehyde as an oxidation product of a primary alcohol. Check that you understand the reactions in these tests.

Requirements

safety spectacles and gloves
6 test-tubes
sodium hydrogensulphite solution, saturated, $NaHSO_3$
ethanal, CH_3CHO -
propanone, CH_3COCH_3 -
2,4-dinitrophenylhydrazine solution, $C_6H_3(NO_2)_2NHNH_2$
sodium hydroxide solution, 2 M NaOH - - - - - - - - - - - - - - - - - - -
Bunsen burner, tripod, gauze and bench mat
beaker, 250 cm³
potassium dichromate(VI) solution, 0.1 M $K_2Cr_2O_7$
sulphuric acid, dilute, 1 M H_2SO_4
Fehling's solutions, 1 and 2 -
silver nitrate solution, 0.05 M $AgNO_3$ - - - - - - - - - - - - - - - -
ammonia solution, 2 M NH_3
iodine solution, 10% (in KI(aq))

Procedure

A. Addition reaction with sodium hydrogensulphite

1. Pour about 2 cm³ of saturated sodium hydrogensulphite solution into
    a test-tube. Point the tube away from you and add, drop by drop, a
    similar volume of ethanal.

2. Shake the tube gently and cool under a stream of cold water. Note
    your observations in a copy of Results Table 89.

3. Repeat steps 1 and 2 using propanone instead of ethanal.

B. Condensation reaction with 2,4-dinitrophenylhydrazine

1. Put 1-2 drops of ethanal in a test-tube and add about 2 cm³ of
    2,4-dinitrophenylhydrazine solution. Note your observations.

2. Repeat for propanone.

C. Reaction with alkali

1. Into a test-tube, pour about 3 cm³ of sodium hydroxide solution and
    10 drops of ethanal.

2. Shake the tube gently and warm in a beaker of hot water for 5-10
    minutes. Note your observations and comment on the smell of
    the product.

3. Repeat for propanone.

D. Oxidation reactions

    (a) With acidified potassium dichromate(VI)

1. Into a test-tube, put 5 drops of ethanal, 2 drops of potassium
    dichromate(VI) solution and 10 drops of dilute sulphuric acid.

2. Shake the tube gently and warm in a beaker of hot water. Note
    your observations.

3. Repeat for propanone.

    (b) With Fehling's solution

1. Into a test-tube, put about 1 cm³ of Fehling's solution 1 and then
    add Fehling's solution 2 dropwise until the precipitate just
    dissolves.

2. Add about seven drops of ethanal. Shake the tube gently and
    place in a beaker of boiling water for five to ten minutes -
    until no further colour change occurs. Note your observations.

3. Repeat for propanone.

(c) With Tollens' reagent

1. Put about 1 cm³ of 0.05 M AgNO₃ into a $\underline{very\ clean}$ test-tube and add three or four drops of sodium hydroxide solution.

2. Drop by drop, add ammonia solution until the precipitate of silver oxide nearly dissolves (do not try to get rid of all the little black specks of silver oxide).

3. Add one or two drops of ethanal, shake the tube gently and place in a beaker of warm water. Note your observations and immediately rinse out the test-tube.

4. Repeat with propanone.

E. Triiodomethane reaction

1. Into a test-tube, place five drops of ethanal followed by 1 cm³ of iodine solution, cork and shake.

2. Drop by drop, add sodium hydroxide solution until the colour of iodine just disappears (about 2 cm³) and a straw-coloured solution remains. Note your observations.

3. Repeat with propanone.

Results Table 89

Test	Observations	
	Ethanal	Propanone
A. Addition reaction with sodium hydrogensulphite		
B. Condensation reaction with 2,4-dinitrophenylhydrazine		
C. Reaction with alkali		

D. Oxidation reactions:
  (a) acidified dichromate(VI)
  (b) Fehling's solution
  (c) Tollens' reagent		

E. Triiodomethane reaction		

Questions

1. Which tests serve to distinguish between ethanal and propanone?

2. Which reaction do you think resulted in the production of a polymer?

3. Which reagent could be used as a general test for a carbonyl compound?

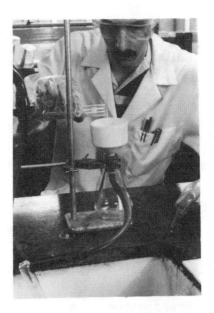

EXPERIMENT 90
Identifying an unknown carbonyl compound

Aim

The purpose of this experiment is to classify
a carbonyl compound by a simple test and
to identify it by the preparation of a
derivative.

Introduction

You are provided with a compound known to be an aldehyde or ketone from a
given list. In the first part of the experiment, you identify the compound as
either an aldehyde or ketone using Tollen's and Fehling's tests. In the
second part, you prepare a derivative of the compound with 2,4-dinitrophenyl-
hydrazine. Finally, you determine the melting point of the derivative in
order to name the particular aldehyde or ketone.

Requirements

A. safety spectacles and protective gloves
   2 test-tubes
   teat-pipette
   silver nitrate solution, 0.05 M $AgNO_3$
   sodium hydroxide solution, 2 M NaOH – – – – – – – – – – – – – – – – – –
   ammonia solution, 2 M NH_3
   Fehling's solutions 1 and 2 – – – – – – – – – – – – – – – – – –
   unknown carbonyl compound, X –
   beaker, 250 cm^3
   Bunsen burner, tripod, gauze and bench mat

B. beaker, 100 cm^3 (or boiling-tube)
   methanol, CH_3OH –
   2,4-dinitrophenylhydrazine solution
   glass stirring rod
   sulphuric acid, dilute, 1 M H_2SO_4
   retort stand, boss and clamp
   apparatus for suction filtration (see Fig. 49)
   spatula
   beaker, 150 cm^3
   steam-bath or 250 cm^3 beaker
   ethanol –
   ice
   filter paper

C. melting-point tubes (at least 2)
   watch-glass
   rubber ring or band
   thermometer, 0-360 °C, long stem
   boiling-tube, fitted with cork and stirrer (see Fig. 50)
   dibutyl phthalate

Procedure

A. Identification of the carbonyl compound as an aldehyde or ketone.

Carry out Tollens' and Fehling's tests on samples of the unknown compound.
Try to remember the procedures before checking (Experiment 89).
Classify the compound as an aldehyde or a ketone.

B. Preparation of a crystalline derivative.

1. Into a 100 cm³ beaker (or boiling-tube) put 0.5 cm³ (10 drops) of the
    unknown compound. (If the substance is solid dissolve 0.5 g in a
    minimum amount of methanol.) Add 5 cm³ of the 2,4-dinitrophenyl-
    hydrazine solution and stir.

2. If precipitation does not occur, carefully add 1 cm³ of dilute
    sulphuric acid.

3. Using the suction filtration
    apparatus in Fig. 49, filter
    the precipitate.

4. Stop suction, either by lifting
    the funnel or by disconnecting
    the tubing, and soak the preci-
    pitate in about 1 cm³ of methanol.
    (If you turn off the tap, you may
    get a 'suck-back' of water.)

5. Resume suction and dry the crystals
    by drawing air through them for a
    few minutes.

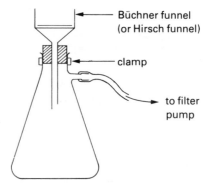

Fig. 49. Suction filtration

6. Recrystallize the solid using the following procedure.

    (a) Transfer the crystals to a 150 cm³ beaker standing on a steam
         bath (or in a 250 cm³ beaker of hot water).

    (b) Dissolve the crystals in the minimum amount of hot ethanol.

    (c) When the crystals have dissolved, cool the solution in an ice-
         water mixture until crystals reappear.

    (d) Filter the crystals as before. If necessary, rinse the beaker
         with the filtrate (not extra solvent) to complete the transfer.
         Finally, wash the crystals with a few drops of cold ethanol.

    (e) Press the crystals thoroughly between two wads of filter paper to
         remove excess solvent. Then put the crystals on another dry
         piece of filter paper placed alongside a Bunsen burner and gauze,
         turning the crystals over occasionally until they appear dry.

C. Determination of the melting-point of the derivative.

1. Take a melting-point tube and push the open end through a pile of the derivative on a watch-glass, until a few crystals have entered the tube.

2. Tap the closed end of the tube vertically against a hard surface, or rub with the milled edge of a coin, to make the solid fall to the bottom.

3. Repeat the filling and tapping procedure until a total length of between 0.5 - 1 cm is compacted at the bottom of the tube. Prepare another tube in this way.

4. Attach one of the prepared melting-point tubes to the thermometer, as shown in Fig. 50.

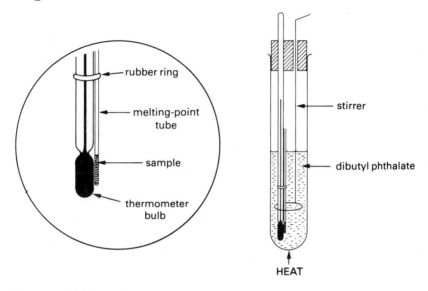

Fig. 50.   Melting-point determination

5. Half-fill the boiling-tube with dibutyl phthalate and position the thermometer with attached tube and the stirrer through the bung, as shown in Fig. 50.

6. Position the apparatus over a micro-burner (or low Bunsen flame) and gauze and gently heat the apparatus, stirring the dibutyl phthalate all the time by moving the stirrer up and down.

7. Keep an eye on the crystals and note the temperature as soon as signs of melting are seen (usually seen as a contraction of the solid followed by a damp appearance). This first reading gives only a rough melting-point but is a guide for the second determination.

8. Remove the burner and the old tube containing derivative. Allow the temperature to drop about 10 °C before positioning a fresh melting-point tube containing another portion of the derivative.

9. Repeat the above procedure in order to obtain a more accurate value of the melting-point. Raise the temperature very slowly (about 2 °C rise per minute) until the crystals melt (take the formation of a visible meniscus as a sign of melting).

10. Compare the melting-point of your crystals with the values given in the table which follows and identify the unknown compound.

11. Check with your teacher or the technician whether you have identified X correctly.

Table 90 Melting-points of some 2,4-dinitrophenylhydrazones

Name	Formula	Boiling point/°C	Melting point of 2,4-dinitrophenyl-hydrazone/°C
Aldehydes			
methanal	HCHO	-21	167
ethanal	CH_3CHO	21	164, 146 (2 forms)
propanal	CH_3CH_2CHO	48	156
butanal	$CH_3CH_2CH_2CHO$	75	123
2-methylpropanal	$(CH_3)_2CHCHO$	64	187
benzaldehyde	C_6H_5CHO	178	237
Ketones			
propanone	CH_3COCH_3	56	128
butanone	$CH_3CH_2COCH_3$	80	115
pentan-2-one	$CH_3CH_2CH_2COCH_3$	102	141
pentan-3-one	$CH_3CH_2COCH_2CH_3$	102	156
hexan-2-one	$CH_3CH_2CH_2CH_2COCH_3$	128	107
4-methylpentan-2-one	$(CH_3)_2CHCH_2COCH_3$	117	95
cyclohexanone	(ring)=O	156	162

Questions

1. What factors decide the choice of solvent in the recrystallization procedure?

2. How were soluble impurities removed from the derivative?

3. In the recrystallization procedure, why were the crystals dissolved in only the minimum amount of ethanol?

4. If your sample had contained insoluble impurities, such as pieces of filter paper, cork, etc., suggest how these might have been removed.

5. Why is it not satisfactory to identify aldehydes and ketones by measuring their boiling-points?

EXPERIMENT 91
Chemical properties of carboxylic acids

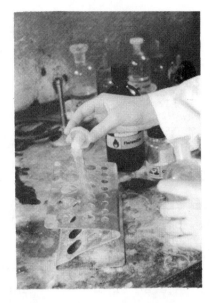

Aim

The purpose of this experiment is to see if carboxylic acids show the typical reactions both of alcohols and of carbonyl compounds.

Introduction

You will be carrying out some test-tube reactions on ethanoic acid. Ethanoic acid melts at 17 °C and therefore freezes in cold weather. Because solid ethanoic acid looks like ice, it is often described as 'glacial'. The reactions to be investigated are:

A   pH of aqueous solution.

B   Reaction with sodium hydrogencarbonate solution.

C   Reaction with sodium.

D   Reaction with phosphorus pentachloride.

E   Reaction with 2,4-dinitrophenylhydrazine.

F   Triiodomethane (iodoform) reaction.

G   Action of iron(III) chloride.

Requirements

safety spectacles and gloves
5 test-tubes in rack
wash-bottle of distilled water
universal indicator solution
ethanoic acid, glacial, CH_3CO_2H — — — — — — — — — — — — — — — 🔥 🧪
sodium hydrogencarbonate solution, saturated, $NaHCO_3$
limewater, $Ca(OH)_2(aq)$
forceps
sodium, Na (1 mm cube under oil) — — — — — — — — — — — — — — ✖
filter papers
wood splint
phosphorus pentachloride, PCl_5 — — — — — — — — — — — — — — — ✖
spatula
ammonia solution, 2 M NH_3
2,4-dinitrophenylhydrazine solution
iodine solution, 10% I_2 in $KI(aq)$
sodium hydroxide solution, 2 M NaOH — — — — — — — — — — — — 🧪
sodium ethanoate (acetate), CH_3CO_2Na
iron(III) chloride solution, 0.1 M $FeCl_3$
Bunsen burner and bench mat

Hazard warning

Glacial ethanoic acid is flammable and corrosive.

Sodium and phosphorus pentachloride are extremely reactive, especially with water. Therefore you MUST:

WEAR SAFETY SPECTACLES AND GLOVES
KEEP STOPPERS ON BOTTLES AS MUCH AS POSSIBLE
KEEP BOTTLES OF FLAMMABLE LIQUIDS AWAY FROM FLAMES
KEEP SODIUM AND PHOSPHORUS PENTACHLORIDE AWAY FROM WATER AND MOIST AIR

Procedure

A. pH of aqueous solution

1. Into a test-tube, pour about 2 cm³ of distilled water and one drop of universal indicator solution. Shake gently.

2. Add a few drops of glacial ethanoic acid, shake gently and note your observations in a copy of Results Table 91.

B. Reaction with sodium hydrogencarbonate solution

Into a test-tube, add about 1 cm³ of sodium hydrogencarbonate solution followed by a few drops of glacial ethanoic acid. Shake gently and test any gas produced.

C. Reaction with sodium (WORK AT A FUME-CUPBOARD, WITH YOUR TEACHER PRESENT)

1. Pour about 2 cm³ of glacial ethanoic acid into a dry test-tube in a rack standing in a fume-cupboard.

2. Using forceps, pick up a 1 mm cube of sodium and blot it free of oil on some filter paper.

3. Drop the clean piece of sodium into the glacial ethanoic acid. Test any gas evolved and note your observations. Do not wash away the mixture until you are sure that the sodium has reacted completely. If some sodium remain unreacted, add a little more glacial ethanoic acid; on no account add water.

D. Reaction with phosphorus pentachloride

1. Pour about 1 cm³ of glacial ethanoic acid into a dry test-tube in a rack standing in a fume-cupboard.

2. A little at a time, carefully add a spatula-measure of phosphorus pentachloride. Bring the wet stopper of an ammonia bottle close to the mouth of the tube and note your observations.

E. Reaction with 2,4-dinitrophenylhydrazine

Into a test-tube, pour about 2 cm³ of 2,4-dinitrophenylhydrazine solution and about 5 drops of glacial ethanoic acid. Shake gently and note your observations.

F. Triiodomethane (iodoform) reaction

Into a test-tube, place ten drops of iodine solution, and five drops of glacial ethanoic acid, followed by sodium hydroxide solution added dropwise until the colour of iodine disappears and a straw-coloured solution remains. Shake the tube gently and note your observations.

G. Reaction with neutral iron(III) chloride solution

Dissolve one or two small crystals of sodium ethanoate in about 2 cm³ of distilled water (this avoids having to neutralize the acid). Add a few drops of iron(III) chloride solution. Shake gently and then heat the solution. Note your observations.

Results Table 91 Reactions of ethanoic acid

Reaction	Observations
A. pH of aqueous solution	
B. Reaction with sodium hydrogencarbonate solution	
C. Reaction with sodium	
D. Reaction with phosphorus pentachloride	
E. Reaction with 2,4-dinitrophenylhydrazine	
F. Triiodomethane reaction	
G. Action of iron(III) chloride	

Questions

1. Does ethanoic acid more closely resemble hydroxy or carbonyl compounds in its reactions? Explain.

2. Which test indicates that ethanoic acid is a stronger acid than phenol?

3. Explain the fact that ethanoic acid is very soluble in water, whereas benzoic acid, $C_6H_5CO_2H$, a solid, is only slightly soluble in water.

EXPERIMENT 92
Identifying salts of carboxylic acids

Aim

The purpose of this experiment is to give you
practice in 'observation and deduction'
exercises, with particular reference to
the salts of carboxylic acids.

Requirements

safety spectacles
protective gloves
6 test-tubes, 2 with corks
test-tube holder and rack
spatula
unknown salts, I and J
wash-bottle of distilled water
sodium hydroxide solution, 2 M NaOH — — — — — — — — — — — — — — — — — —
sulphuric acid, dilute, 1 M H_2SO_4
Bunsen burner and bench mat
wood splint
litmus papers or pH papers
limewater, $Ca(OH)_2(aq)$
test-tube and bung fitted with right-angled delivery tube
2,4-dinitrophenylhydrazine solution, $C_6H_3(NO_2)_2NHNH_2$
sulphuric acid, concentrated, H_2SO_4 — — — — — — — — — — — — — — —
teat-pipette

Hazard warning

Sodium hydroxide solution is corrosive.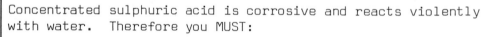

Concentrated sulphuric acid is corrosive and reacts violently
with water. Therefore you MUST:

WEAR SAFETY SPECTACLES AND GLOVES

If contact with skin does occur, wash immediately with plenty of
cold water.

Dispose of residues containing concentrated sulphuric acid by cooling first
and then pouring into plenty of cold, running water.

Procedure

You are provided with two salts, I and J, which are composed of the same elements. Carry out the following tests and record your observations and inferences in (a larger copy of) the table. Then answer the question which follows the table.

(Note that the question does <u>not</u> say anything about carboxylic acids - we have made it easier for you!)

Results Table 92

Test	Observations	Inferences

(a) Dissolve some of I in the minimum quantity of distilled water and use portions of the solution for the following tests.
 	
   (i) To 1-2 cm³ of the solution of I add aqueous sodium hydroxide until in excess.
 	
   (ii) To 1-2 cm³ of the solution of I add an equal volume of dilute sulphuric acid, shake the tube, and allow the solution to stand.		

(b) Add a little dilute sulphuric acid to some solid I and heat the mixture.		
(c) Heat a little solid I in a Pyrex test-tube.		
(d) Using the apparatus shown, heat a little of I in a test-tube and pass the vapour evolved into 2,4-dinitrophenylhydrazine reagent.		
(e) Heat a little solid J in a Pyrex test-tube.		
(f) To a little solid J add a little concentrated sulphuric acid (CARE!) and warm gently.		

Comment on the identify of I and J.

EXPERIMENT 93
Chemical properties of ethanoyl chloride

Aim

The purpose of this experiment is to illustrate the chemistry of carboxylic acid derivatives by investigating the reactions of ethanoyl chloride with some nucleophilic reagents.

Introduction

You or your teacher will be carrying out the following reactions of ethanoyl chloride - a typical acyl halide.

A. Reaction with water.   C. Reaction with ammonia.

B. Reaction with ethanol.   D. Reaction with phenylamine.

Since some of the reactions are violent, your teacher may decide to demonstrate this experiment. If you do it yourself, you must take great CARE!

Requirements

safety spectacles
protective gloves
4 beakers, 100 cm³
wash-bottle of distilled water
4 teat-pipettes
ethanoyl chloride, CH_3COCl — — — — — — — — — — — — — — — — — — ✖
ammonia solution, dilute, 2 M NH_3
glass rod
universal indicator paper
iron(III) chloride solution, 0.1 M $FeCl_3$
test-tube
sodium carbonate solution, 1 M Na_2CO_3
ethanol, C_2H_5OH— — — — — — — — — — — — — — — — — — — 🔥

ammonia solution, concentrated, '0.880' NH_3 — — — — — — — — — — — — 🧪

phenylamine, $C_6H_5NH_2$ — — — — — — — — — — — — — — — — — ☠

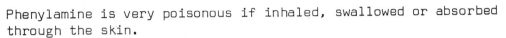

Hazard warning

Ethanoyl chloride gives off a vapour which burns the skin, eyes and respiratory tract. Some of its reactions are violent.

Phenylamine is very poisonous if inhaled, swallowed or absorbed through the skin.

Concentrated aqueous ammonia burns the skin and the vapour irritates the eyes. Therefore you MUST:

WEAR SAFETY SPECTACLES AND GLOVES.
KEEP STOPPERS ON THE BOTTLES AS MUCH AS POSSIBLE.
PERFORM THE EXPERIMENT IN A FUME CUPBOARD PROTECTED BY A SAFETY GLASS.

Procedure

A. Reaction with water

1. Pour about 5 cm³ of distilled water into a small beaker placed in the fume cupboard with the front pulled down as far as is practicable.

2. Wearing safety spectacles, add a few drops of ethanoyl chloride to the water. Bring the moist stopper of a bottle of ammonia solution near to the top of the beaker and note your observations.

3. Neutralize the solution in the beaker by adding dilute aqueous ammonia dropwise until a drop of the solution from a glass rod gives a neutral colour to universal indicator.

4. In a test-tube, neutralize about 1 cm³ of iron(III) chloride solution by adding sodium carbonate solution dropwise until a faint precipitate just remains on shaking.

5. Add a few drops of the neutral iron(III) chloride to the neutral solution prepared in step 3 and note your observations in a copy of Results Table 93.

B. Reaction with ethanol

1. Repeat steps 1 and 2 of Experiment A using ethanol instead of water in the beaker.

2. Add sodium carbonate solution until there is no further effervescence and note the smell of the product.

C. Reaction with ammonia

Repeat steps 1 and 2 of Experiment A using concentrated aqueous ammonia instead of water in the beaker. Take special care when you add the ethanoyl chloride a drop at a time. Note your observations.

D. Reaction with phenylamine

1. Pour five drops of phenylamine in a small beaker placed in the fume cupboard with the front pulled down as far as is practicable.

2. Wearing safety spectacles, add a few drops of ethanoyl chloride to the phenylamine. Note your observations.

Results Table 93

Reaction	Observations

A. Reaction with water
    Product tested with ammonia
    Product tested with iron(III) chloride	

B. Reaction with ethanol
    Product tested with ammonia
    Smell of product	

C. Reaction with ammonia solution	
D. Reaction with phenylamine	

Questions

1. Using the general equation for the nucleophilic substitution reactions of carboxylic acid derivatives shown below

$$R-\underset{\underset{O}{\|}}{C}-Y \ + \ HZ \ \longrightarrow \ R-\underset{\underset{O}{\|}}{C}-Z \ + \ HY \quad (HZ = \text{nucleophilic reagent})$$

together with your observations from the experiment, write equations for the reactions of ethanoyl chloride and the following nucleophilic reagents. Name the products.

(a) Water, H_2O (c) Ammonia, NH_3

(b) Ethanol, C_2H_5OH (d) Phenylamine, $C_6H_5NH_2$

2. Explain the order of reactivity of the four nucleophiles towards ethanoyl chloride.

EXPERIMENT 94
Preparing an ester

Aim

The purpose of this experiment is to prepare
a sample of phenyl benzoate, purify it by
recrystallization, and measure its melting-
point.

Introduction

You prepare phenyl benzoate by shaking phenol with benzoyl chloride in
alkaline solution.

$$C_6H_5COCl + C_6H_5OH \rightarrow C_6H_5CO_2C_6H_5 + HCl$$

The product appears as a solid and you purify it by filtering, dissolving in
hot ethanol, and cooling to recrystallize the ester.

Requirements

A. safety spectacles and protective gloves
    weighing-bottle
    spatula
    phenol, C_6H_5OH —
    access to balance, sensitivity ± 0.01 g
    conical flask, 250 cm³, with tight-fitting rubber bung
    measuring cylinder, 100 cm³
    sodium hydroxide solution, 2 M NaOH — — — — — — — — — — — — — —
    measuring cylinder, 10 cm³
    benzoyl chloride, C_6H_5COCl — — — — — — — — — — — — — — — — |
    suction filtration apparatus as in Fig. 51
    wash-bottle of distilled water

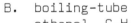

B. boiling-tube
    ethanol, C_2H_5OH
    glass rod
    water-bath or 250 cm³ beaker
    Bunsen burner, tripod, gauze and bench mat
    thermometer, 0-100 °C
    ice
    suction filtration apparatus as in Fig. 51
    filter papers
    specimen bottle
    access to balance, sensitivity ± 0.01 g

C. melting-point tubes (at least 2)
    watch-glass
    rubber ring or band
    thermometer, 0-100 °C, long stem
    boiling-tube fitted with cork and stirrer (see Fig. 50, Experiment 90)
    dibutyl phthalate
    retort stand, boss and clamp

Procedure

A. Preparation of phenyl benzoate

1. Transfer about 5.0 g of phenol into a weighing-bottle and weigh it to the nearest 0.01 g.

2. Into a conical flask pour 90 cm³ of 2 M sodium hydroxide and the bulk of the phenol from the weighing-bottle.

3. Reweigh the weighing-bottle, with any remaining phenol, to the nearest 0.01 g.

4. In a fume cupboard pour 9 cm³ of benzoyl chloride into the conical flask.

5. Insert the bung securely and shake the bottle for 15 minutes, carefully releasing the pressure every few minutes as the flask gets warm.

6. Cool the flask under cold, running tap-water.

7. Filter the crude product using a suction filtration apparatus (Fig. 51). Use a spatula to break up the lumps of ester on the filter paper, being careful not to puncture the filter paper.

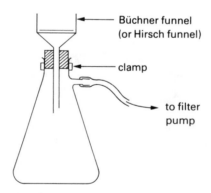

Fig.51. Suction filtration

8. Pour more water over the crude ester to destroy any remaining benzoyl chloride.

B. Recrystallization

1. Transfer the crystals to a boiling-tube and just cover them with ethanol.

2. Place the beaker in a water-bath or beaker of hot water, kept at about 60 °C and stir with a glass rod.

3. If some solid ester is still visible, add just enough ethanol to dissolve it completely after stirring.

4. In order to allow the separation of the ester as a solid rather than an oily liquid (phenyl benzoate has a low melting-point) add more ethanol to double the volume of solution.

5. Cool the solution in an ice-water mixture until crystals appear.

6. Filter the crystals through the suction apparatus, using a clean Büchner funnel and filter paper. To avoid losing any solid, break the vacuum and use the filtrate to rinse the boiling-tube into the funnel.

7. Using suction again, rinse the crystals with about 1 cm³ of cold ethanol and drain thoroughly.

8. Press the crystals between two wads of filter paper to remove excess solvent. Then put the crystals on another dry piece of filter paper placed alongside a Bunsen burner and gauze, turning the crystals over occasionally until they appear dry. Don't let them get too hot or they will melt!

9. Weigh the dry crystals in a pre-weighed specimen bottle and record the mass of your sample of phenyl benzoate.

C. Determination of melting-point

Follow the procedure which is described in Experiment 90 and record the melting-point of your sample of phenyl benzoate.

Results Table 94

Mass of weighing-bottle + phenol	g
Mass of weighing-bottle after emptying	g
Mass of phenol	g
Mass of specimen bottle	g
Mass of specimen bottle + phenyl benzoate	g
Mass of phenyl benzoate	g
Melting-point of phenyl benzoate	°C

Questions

1. Calculate the maximum mass of phenyl benzoate that could be formed from the mass of phenol you used, assuming complete conversion according to the equation.

2. Calculate the percentage yield of phenyl benzoate using the expression:

$$\frac{\text{Mass obtained}}{\text{Theoretical maximum mass}} \times 100 = \% \text{ yield}$$

3. The melting-point of pure phenyl benzoate is 69 °C. What can you say about the purity of your sample?

EXPERIMENT 95
Observation and deduction exercise

Aim

This experiment is intended primarily for students preparing for an A-level practical examination in which one of the questions could require the identification of an unknown compound by means of specified tests.

Introduction

The procedure which follows is taken from three A-level practical examination papers; read it carefully and report fully.

Requirements

safety spectacles and gloves
solution labelled F
6 test-tubes in a rack
teat-pipette and test-tube holder
boiling-tube and spatula
2,4-dinitrophenylhydrazine solution
silver nitrate solution, 0.05 M $AgNO_3$
sodium hydroxide solution, 2 M NaOH
ammonia solution, 2 M NH_3
limewater, saturated, $Ca(OH)_2$
wash-bottle of distilled water
sodium hydrogencarbonate, $NaHCO_3$
potassium iodide solution, 0.6 M KI
sodium chlorate(I) (hypochlorite) solution, NaClO
liquid labelled G
crucible lid
Bunsen burner, tripod, gauze and mat
2 wood splints
2 beakers, 100 cm^3
beaker, 250 cm^3
sodium hydroxide solution, 6 M NaOH
stirring rod
hydrochloric acid, concentrated, HCl
liquid labelled H
anhydrous sodium carbonate, Na_2CO_3
phosphorus pentachloride, PCl_5
ethanoic (acetic) acid, glacial CH_3CO_2H
phenylamine (aniline), $C_6H_5NH_2$

Hazard warning

Concentrated hydrochloric acid, phosphorus pentachloride, unknown compound H and solutions of silver nitrate, sodium chlorate(I) (sodium hypochlorite) and sodium hydroxide are all corrosive. Phenylamine (aniline) and 2,4-dinitrophenylhydrazine are toxic by skin absorption. Therefore you MUST:

WEAR SAFETY SPECTACLES AND GLOVES THROUGHOUT.

Phosphorus pentachloride and compounds G and H produce harmful vapours. Therefore you MUST:

PERFORM EXPERIMENTS WITH THESE COMPOUNDS AT A FUME CUPBOARD.
KEEP TOPS ON BOTTLES AS MUCH AS POSSIBLE.

Compounds G and H are flammable. Therefore you MUST:

KEEP THESE COMPOUNDS AWAY FROM FLAMES.

Procedure

You are provided with three organic compounds, labelled F, G and H. Each compound contains the elements carbon, hydrogen and oxygen only. Carry out the following experiments. Record your observations and inferences in (a larger copy) of Results Table 95. Comment on the types of chemical reaction occurring and, where possible, deduce the functional groups present in these compounds. Then answer the questions which follow.

(1) Comment on the structural features of F.

(2) Comment on the structural features of G.

(3) What functional group is present in H? Give your reasoning.

Results Table 95

EXPERIMENT	OBSERVATIONS	INFERENCES
(a) To 2 cm³ of the solution of F add 2, 4-dinitrophenylhydrazine reagent.		
(b) Prepare a sample of Tollens' reagent as follows: to 5 cm³ of aqueous silver nitrate in a test-tube add 1-2 drops of aqueous sodium hydroxide. Then add dilute aqueous ammonia until only a trace of precipitate remains. Now add 5 drops of the solution of F and place the tube in hot water. (Pour the contents of the tube down the sink on completion of the test.)		
(c) Add a little of the solution of F to some sodium hydrogencarbonate.		
(d) To 1 cm³ of the solution of F add 3 cm³ of aqueous potassium iodide and then 10 cm³ of sodium chlorate(I) (sodium hypochlorite) solution.		
(e) Place ONE drop of G on an inverted crucible lid and ignite G from above.		
(f) To 1-2 cm³ of G add 2,4-dinitrophenylhydrazine reagent.		
(g) Prepare a sample of Tollens' reagent as in (b). Add 5 drops of G, shake the mixture and place the tube in hot water. (N.B. Pour the contents of the tube down the sink on completion of the test.)		
(h) To 0.5 cm³ of G add 2 cm³ of 6 M sodium hydroxide. Warm the mixture and stir well for 5 minutes. Then add sufficient water to dissolve any residue which has formed. Separate a portion of the aqueous layer and add a little concentrated hydrochloric acid to it. Cool.		

(i) Test to see if H is miscible with
  (i) cold water,
  (ii) hot water.		

(j) (i) To about 2 cm³ of H in a dry beaker add a little anhydrous sodium carbonate. (ii) Now add about 5 cm³ of hot water to the mixture from test (j),(i).		
(k) (This reaction should be performed in a fume cupboard.) Place 2 or 3 cm³ of H in a dry beaker. Add a little phosphorus pentachloride (CARE!).		
(l) Place 2 cm³ of glacial ethanoic acid and 2 cm³ of H in a boiling tube and add 2 cm³ of phenylamine (aniline). Heat cautiously until the mixture begins to boil. Remove from the flame and allow to stand for 2 minutes. Now pour the mixture into about 125 cm³ of cold water with stirring and allow to stand.		

EXPERIMENT 96
The glycine/copper(II) complex

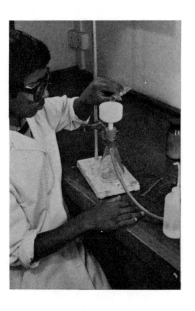

Aim and Introduction

The purpose of this simple experiment is to prepare a complex from glycine and copper(II) ions.

Requirements

safety spectacles
wash-bottle of distilled water
test-tube
spatula
glycine, $CH_2NH_2CO_2H$
copper(II) carbonate, $CuCO_3$ (powdered)
stirring rod
Büchner funnel (small) or Hirsch funnel
filter paper
filter tube with side-arm
filter pump and pressure tubing
crystallizing dish

Procedure

1. To about 10 cm³ of distilled water in a test-tube, add about 0.5 g (small spatula-measure) of glycine. Note how easily it dissolves.

2. Slowly add powdered copper(II) carbonate, stirring the contents of the tube between additions. Keep adding the powder until it is in excess.

3. Set up a small Büchner funnel or Hirsch funnel in a side-arm filter tube.

4. Transfer the filtrate to a crystallizing dish and allow it to stand. Note the colour of the solution and whether crystals are formed.

Questions

1. Copper(II) carbonate is only slightly soluble but during the experiment some CO_3^{2-} ions do dissolve. Would you expect this to make the solution slightly acidic, neutral or alkaline? Explain.

2. Bearing in mind your answer to Question 1, what form of glycine would you expect to predominate in the solution?

3. How many bonds is each glycine molecule capable of making with Cu^{2+}(aq) under these conditions?

EXPERIMENT 97
The biuret test for proteins

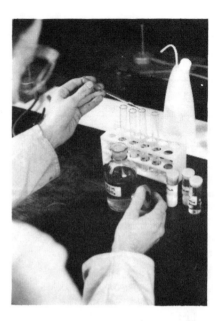

Aim

This experiment is intended to give you practical experience of the biuret test for proteins.

Introduction

The biuret test is based on a coloured complex formed between copper(II) ions and the peptide links of neighbouring protein chains. You carry out the test on examples of available protein material.

Requirements

safety spectacles
4 test-tubes in a rack
spatula and teat-pipette
protein sample(s) (e.g. egg albumin, gelatin, fresh milk)
wash-bottle of distilled water
sodium hydroxide solution, 2 M NaOH — — — — — — — — — — — — — — —
copper(II) sulphate solution, 0.2 M $CuSO_4$

Procedure

1. Take a small amount (enough to cover a spatula tip if solid, about 2 cm³ in the bottom of a test-tube, if liquid) of one of the proteins. Dissolve it in water so that the total volume is no more than ¼ of a test-tube. If necessary, warm the tube and then allow its contents to cool.

2. Add an equal volume of 2 M sodium hydroxide solution followed by 5 drops of 0.2 M copper(II) sulphate solution.

3. Leave the tube to stand, if necessary, and note the colour that develops.

4. Repeat steps 2 and 3 using water instead of protein solution. Compare the colour of this 'blank' tube with the colour of your protein solution from 2. If you don't get a definite colour, try again with a more concentrated protein solution in step 1.

5. Repeat steps 1 to 3 using other protein(s).

Questions

1. How do you think the biuret test might be used to estimate the concentration of a protein in a solution?

2. The biuret test is named after the compound:

$$H_2N-\overset{\overset{\displaystyle O}{\|}}{C}-NH-\overset{\overset{\displaystyle O}{\|}}{C}-NH_2$$
biuret

   Give the fact that the Cu^{2+} ion forms 4-coordinate, planar complexes, suggest a formula for the complex between Cu^{2+} and two protein chains.

3. If you hydrolyzed a protein and converted it completely into its constituent amino acids, would you expect to get a positive biuret test? Explain.

EXPERIMENT 98
Paper chromatography

Aim

The purpose of this experiment is to
illustrate the use of paper chromatography
for the separation and identification of
amino acids.

Introduction

In this experiment you separate a mixture of three amino acids by means of
paper chromatography. From the chromatogram you calculate R_f values for the
individual amino acids.

Requirements

safety spectacles and protective gloves
access to fume cupboard
measuring cylinder, 10 cm^3
beaker, 400 cm^3, tall-form
watch-glass (big enough to cover beaker)
ethanol, C_2H_5OH
wash-bottle of distilled water
ammonia solution, 0.880 NH_3 — — — — — — — — — — — — — — — — — — —
square of chromatography paper, 12.5 cm x 12.5 cm
pencil, ruler and 2 paper clips
4 melting-point tubes
aspartic acid solution, 0.01 M
leucine solution, 0.01 M
lysine solution, 0.01 M
mixture of the three amino acids above
2 retort stands, bosses and clamps
hair-dryer
ninhydrin aerosol spray —
oven (105 °C)

Hazard warning

Concentrated ammonia solution is corrosive and gives off harmful
vapour. Therefore you MUST:

WEAR SAFETY SPECTACLES AND GLOVES
KEEP BOTTLES CLOSED AS MUCH AS POSSIBLE

Ninhydrin sprays give off toxic fumes. Therefore you MUST:

WORK AT A FUME CUPBOARD
READ THE INSTRUCTIONS ON THE AEROSOL CAN

247

Procedure

1. In a fume cupboard, prepare the solvent mixture by pouring the following into a 400 cm³ tall-form beaker:

    24 cm³ of ethanol,
    3 cm³ of distilled water,
    3 cm³ of 0.880 ammonia.

2. Cover the beaker with a watch-glass. Swirl to mix the liquids and leave to stand.

3. Handling it only by the top edge, place a square of chromatography paper on a clean sheet of file paper. With a pencil (not a pen) draw lines and labels as shown in Fig. 52.

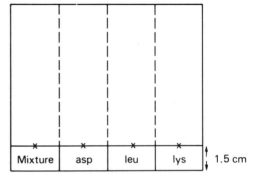

Fig. 52.

4. Dip a clean melting-point tube into the solution of mixed amino acids and then touch it briefly on the appropriate labelled cross so that a spot, no more than 5 mm across, appears on the paper.

5. Using a fresh tube each time, repeat step 4 for each of the three solutions of single amino acids.

6. Place a clean ruler with its edge along one of the dashed lines and hold it firmly in place with one hand. Without touching it with your fingers, fold the chromatography paper along the line by sliding your hand under the file paper and lifting.

7. Repeat the folding procedure for the other two lines so that the opposite edges of the paper almost meet to form a square cross-section.

8. Hold the paper by the edge furthest from the start line, and place it in the beaker so that it does not touch the sides. Replace the cover and leave to stand.

9. Clamp the ruler horizontally at a height of 20-30 cm between two retort stands in a fume cupboard. This is to support the chromatography paper for drying when the run has finished.

10. While you are waiting, get on with some other work, but look at the paper every 10 minutes to see how far the solvent has soaked up the paper.

11. When the solvent has reached nearly to the top of the paper (30-40 minutes) or when you have only 15 minutes laboratory time left, whichever is the sooner, remove the paper from the beaker, open it out and clip it on to the ruler to dry. You can hasten the drying with a hair-dryer.

12. When the paper is dry, spray it evenly with ninhydrin solution. Dry it again and then heat it in an oven at 105 °C for 5 minutes.

13. Remove the paper from the oven and mark with a pencil the positions of the coloured spots.

14. Measure the distances from the origin line to the centres of the spots and record them in a copy of Results Table 98.

Results and calculations

Calculate an R_f value for each spot as follows:

$$R_f = \frac{\text{distance travelled by spot}}{\text{distance travelled by solvent}}$$

Results Table 98

| Amino-acid | Distances travelled/cm | | R_f value |
	by solvent	by amino-acid	
Aspartic acid - alone			
- in mixture			
Leucine - alone			
- in mixture			
Lysine - alone			
- in mixture			

Questions

1. For each amino-acid, compare your two R_f values with each other and with our specimen results which you can obtain from your teacher. Why do you think there is some variation?

2. Why do R_f values change when a different solvent is used?

3. Why is it so important to avoid touching the chromatography paper with your fingers?

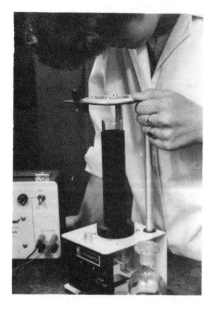

EXPERIMENT 99
Reactions of carbohydrates

Aim

The purpose of this experiment is to illustrate the reactions of some sugars and the
polysaccharide starch.

Introduction

This experiment is divided into five parts:

A. Dehydration. Here you examine the effect of heat and of concentrated
   sulphuric acid on sugars.

B. Oxidation. You use Fehling's and Tollens' reagents to test sugars for
   reducing power.

C. Carbonyl derivatives. You make the 2,4-dinitrophenylhydrazone and the
   phenylhydrazone of a sugar.

D. Hydrolysis. You examine the effects of hot dilute acid and saliva on
   sucrose and starch.

E. Polarimetry. You use a polarimeter to observe the rotation of plane-
   polarized light by sugars and to follow the hydrolysis of sucrose.

General requirements for Parts A, B, C and D

safety spectacles
6 test-tubes in a rack
beaker, 250 cm^3 (for use as a water-bath)
Bunsen burner, tripod, gauze and bench mat
wash-bottle of distilled water
spatula
glucose, $C_6H_{12}O_6$
fructose, $C_6H_{12}O_6$
sucrose, $C_{12}H_{22}O_{11}$
maltose, $C_{12}H_{22}O_{11}$
starch, $(C_6H_{11}O_5)n$
sulphuric acid, dilute, 1 M H_2SO_4

Part A

protective gloves
sulphuric acid, concentrated, H_2SO_4
wood splint
limewater, $Ca(OH)_2(aq)$
potassium dichromate(VI) solution, 0.1 M $K_2Cr_2O_7$
filter paper strips

Part B

Fehling's solutions 1 and 2
ammonia solution, 2 M NH_3
silver nitrate solution, 0.05 M $AgNO_3$

Part C

2,4-dinitrophenylhydrazine solution
phenylhydrazinium chloride, $C_6H_5NHNH_3Cl$
sodium ethanoate, hydrated, $CH_3CO_2Na \cdot 3H_2O$

Part D

boiling water-bath
ammonia solution, 2 M NH_3
litmus paper
Fehling's solution or Tollens' reagent as in part B
iodine solution, 0.1 M I_2 in KI(aq)

Part E

simple polarimeter

Hazard warning

Concentrated sulphuric acid is very corrosive and reacts violently
with water, especially when hot. Therefore you MUST:

WEAR SAFETY SPECTACLES AND PROTECTIVE GLOVES,
USE SMALL QUANTITIES AND COOL RESIDUES BEFORE DISPOSAL,
POUR COLD RESIDUES SLOWLY INTO PLENTY OF COLD WATER, STIRRING TO DISPERSE HEAT

Fehling's solution 2 is corrosive because it contains sodium hydroxide
It is likely to spurt out of a test-tube during heating.
Therefore you MUST:

WEAR SAFETY SPECTACLES
ENSURE THAT NOBODY IS IN LINE WITH A TEST-TUBE DURING HEATING.

Tollens' reagent can become explosive if allowed to evaporate to
dryness. Therefore you MUST:

WASH AWAY RESIDUES WITH PLENTY OF WATER

Procedure

A. Dehydration

1. Warm about 0.5 g of glucose or sucrose in a dry test-tube. Use a low
    flame and heat the tube gently. Record your observations in a copy of
    Results Table 99a, noting particularly any changes in state, colour,
    viscosity and smell.

2. Carefully add about 1 cm³ of concentrated sulphuric acid to about
    0.5 g of glucose or sucrose in a test-tube.

3. Warm the mixture gently and then remove the tube from the flame to
    observe and record the changes which occur without further heating.

4. Heat the mixture more strongly and test for carbon monoxide, carbon
    dioxide and sulphur dioxide.

B. Oxidation

1. Dissolve about 0.1 g of glucose in 2 cm³ distilled water in a test-tube.

2. Add 1 cm³ each of Fehling's solutions 1 and 2. Heat the tube
    carefully to keep the mixture <u>just</u> boiling for about 30 seconds.

3. Note the colour of the solution and whether any precipitate is formed.

4. Repeat steps 1 to 3 using fructose, maltose and sucrose.

5. Prepare some ammoniacal silver nitrate solution (Tollen's reagent) for your own use as follows. Add ammonia solution drop by drop to about 5 cm³ of silver nitrate solution until the resulting buff precipitate <u>almost</u> redissolves on shaking.

6. Dissolve about 0.1 g of glucose in 2 cm³ distilled water in a <u>clean</u> test-tube.

7. Add about 2 cm³ of Tollens' reagent and heat the tube in a beaker of boiling water. Note any colour change that takes place.

8. Repeat steps 5 to 7 with fructose, maltose and sucrose.

C. <u>Carbonyl derivatives</u>

1. Dissolve about 0.5 g of glucose in 1 cm³ distilled water and add 5 cm³ of 2,4-dinitrophenylhydrazine solution.

2. If crystals do not form, add a little dilute sulphuric acid, warm the test-tube and then cool under running cold water.

3. Weigh out 0.2 g of glucose or fructose, 0.4 g phenylhydrazinium chloride and 0.6 g hydrated sodium ethanoate and transfer into a clean, dry test-tube.

4. Add 4 cm³ distilled water.

5. Note the time and heat in a boiling water-bath, shaking the tube occasionally.

6. Record the time when crystals first appear.

7. If you have time, ask for extra apparatus so that you can filter the crystals, wash with a little distilled water, dry between filter papers and recrystallize from ethanol.

8. Measure the melting-point of a dried sample of the purified crystals and compare it with the value given in tables of melting-points.

D. (a) <u>Hydrolysis of sucrose</u>

1. Take about 0.3 g sucrose, add 4 cm³ distilled water and shake to dissolve. Add 1 cm³ of dilute sulphuric acid.

2. Heat the tube in a boiling water-bath for 5 minutes.

3. Add enough dilute aqueous ammonia to neutralize the solution.

4. Carry out a test with either Fehling's solution or Tollens' reagent to see whether you can detect any reducing sugar.

D. (b) <u>Hydrolysis of starch</u>

1. Place about 2 cm³ of starch solution in each of four test-tubes labelled A, B and C and D.

2. Add a little saliva solution to each of tubes A and B.

3. Boil the contents of tube A for 2 or 3 minutes, then place both tubes A and B in a water bath at 40 °C for 20 minutes.

4. Add 1 cm³ dilute sulphuric acid to C and place it in a boiling water-bath for 15 minutes.

5. Neutralize solution C with dilute aqueous ammonia.

6. Divide the contents of each of the four tubes into two parts. Add two or three drops of iodine solution to one part and test the other with Fehling's solution or Tollens' reagent as in part B of this experiment.

Results Table 99a

Experiment	Observations				
	Glucose	Fructose	Sucrose	Maltose	Starch

A. Dehydration
    (a) Action of heat
    (b) Sulphuric acid
		▓		▓	▓

B. Oxidation
    (a) Fehling's soln.
    (b) Tollen's reagent
					▓

C. Carbonyl derivatives
    (a) 2,4-dinitro-
         phenylhydrazone
    (b) Phenylhydrazone
			▓	▓	▓

D. Hydrolysis	▓	▓	▓	▓	

E. Polarimetry

We assume that
you will use a
polarimeter
like this one
shown in
Fig. 53.
If this is not
the case, ask
your teacher
to modify our
instructions.

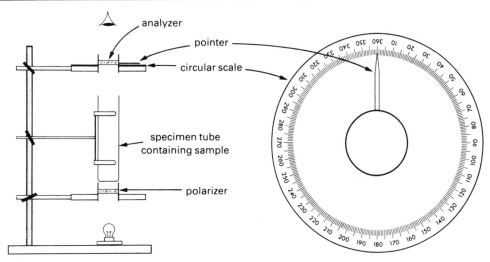

Fig. 53. A simple polarimeter

1. Remove the specimen tube and look vertically down at the light source
    through both polaroid films. If possible, insert a filter to limit the
    light to a narrow band of wavelengths.

2. Rotate the analyzer until you find a position which allows no light (or
    hardly any) to pass, and set the pointer to zero.

3. Fill the specimen tube with a fairly concentrated solution of glucose.

4. Place the specimen tube in position. Note, in a copy of Results Table 99b,
    the new setting of the analyzer which extinguishes the light.

5. Halve the light path in the liquid by pouring away half of the solution.
    Note the new setting of the analyzer which extinguishes the light.

6. Refill the specimen tube by adding distilled water. Again, adjust the
    analyzer and note the new setting.

If you have time, you may like to try another short experiment.

1. Dissolve 50 g of sucrose in 50 cm^3 of hot water and leave to cool.

2. Add 20 cm^3 of concentrated hydrochloric acid, mix well and pour into the
    polarimeter tube.

3. Take a reading, α_t, of the setting of the analyzer and note the time, t. Record your results in a copy of Results Table 99b.

4. Take further readings at intervals as shown until there is no further change. 60 minutes should be enough for α_∞.

Results Table 99b

Glucose	Movement of analyzer from zero								
	Initial	½ volume	Diluted						
Sucrose	Time, t/min			0	3	6	10	15	20	30	∞
	Analyzer reading, $\alpha_t/°$									
	$\alpha_t - \alpha_\infty/°$									

+ indicates clockwise

- indicates anticlockwise

Questions

A. 1. What are the main products when concentrated sulphuric acid reacts with glucose?

2. Why is this reaction called a dehydration?

3. What happens to the sulphuric acid during the reaction?

B. 4. Name the products formed when Tollens' reagent and Fehling's solution react with glucose.

5. Explain why glucose and fructose, but not sucrose, reduce silver(I) and copper(II) in solution.

C. 6. Give the common name and formula of the product formed when glucose or fructose reacts with phenylhydrazine. (The same product in each case.)

D. 7. Write an equation for the hydrolysis of sucrose.

8. Why is this reaction known as the <u>inversion</u> of sucrose?

9. Name the enzyme, found in yeast, which catalyses this reaction.

E. 10. Calculate the concentration of the solution of glucose which you used, given the following information.

The rotation caused by a 10 cm* column of solution at a concentration of 1 g cm^{-3}* is constant for a given substance and is known as the specific rotation, $[\alpha]$. $[\alpha]$ (D-glucose) = +52.5° cm^3 g^{-1} dm^{-1}

Specific rotation is related to observed rotation, α, by the expression:

$$[\alpha] = \frac{\alpha}{lc}$$

l (in dm*) is the length of the light path in the solution, and c is the concentration (g cm^{-3}*)

*Note the unusual units.

(Strictly speaking, the wavelength of the polarized light and the temperature should also be constant at specified values, but you may ignore these for an approximate calculation.)

11. If you did the hydrolysis of sucrose experiment with the polarimeter, plot a graph of $(\alpha_t - \alpha_\infty)$ against t. Comment on its shape.